Luana Nepomuceno Gondim Costa Lima
Cristiane Cunha Frota
Lígia Sansigolo Keer

Molecular epidemiological study of Leprosy in Fortaleza, Ceará

Luana Nepomuceno Gondim Costa Lima
Cristiane Cunha Frota
Lígia Sansigolo Keer

Molecular epidemiological study of Leprosy in Fortaleza, Ceará

Molecular epidemiology of leprosy in Fortaleza, Ceará

ScienciaScripts

Imprint

Cover image: www.ingimage.com

This book is a translation from the original published under ISBN 978-613-9-61980-1.

Publisher:
Sciencia Scripts
is a trademark of
Dodo Books Indian Ocean Ltd. and OmniScriptum S.R.L publishing group

120 High Road, East Finchley, London, N2 9ED, United Kingdom
Str. Armeneasca 28/1, office 1, Chisinau MD-2012, Republic of Moldova, Europe
Printed at: see last page
ISBN: 978-620-7-71079-9

SUMMARY

I dedicate this thesis to two very important people in my life. My mother, Silvana, who provided everything I needed to create an environment conducive to making my dreams come true. My brother, Tibério, my friend and companion since childhood, with whom I have an indestructible bond of love.

THANKS

To my mother, to whom, in addition to dedicating this book, I would like to thank for all her support and encouragement, for her strength in the most complicated moments, for her understanding, love and for her example of overcoming, of being a person, a woman and a mother. For being my eternal mentor in life's most complex issues and for making my dreams and achievements possible. To my brother who is part of who I am, thank you for your affection and for your very existence.

A special thanks to my grandfather Antônio Simoes Gondim, who was always present in my life, contributing to my education and growth as a father and grandfather, and whom I love very much.

To my grandmother, Luciola, for all her dedication and love for my family, who during her earthly existence was always by my side and who, although she is no longer with me on this plane, I know is very proud.

To my stepfather, Raimundo Nonato Franco, for being part of my life, for his warmth and the support of a father.

To Prof. Dr. Ligia Kerr, who opened the door to the field of Leprosy studies, thank you very much for your welcome, attention, friendship, affection, patience and guidance over all these years. For your constant presence in my academic life and for your example of persistence, determination, courage and competence.

To Prof. Dr. Cristiane Cunha Frota, who added so much knowledge and experience to my way of seeing the world and acting in it, particularly in relation to research. For being a reference for me as a person and as a professional, for her friendship, patience and guidance in this and other work, which began in 2002 at the start of my degree.

To the Graduate Program in Medical Sciences, through the professors and staff, for the training and support during my doctorate.

To Dr. Rosa Maria Salani Mota for her collaboration in the statistical analyses and to D.". Rosa Livia Freitas de Almeida for her help with georeferencing.

To the team at the Dona Libânia National Reference Center for Health Dermatology for the samples provided and for supporting the development of the project.

To CAPES and CNPq for their financial support.

SUMMARY

The study of *M. leprae* DNA detection in nasal secretions of individuals and *M. leprae* genotyping methodologies have complemented the epidemiology of leprosy. This study evaluated the positivity of *M. leprae* DNA in the nasal secretion (NS) of patients with leprosy and healthy individuals; and the genetic variability between strains of *M. leprae,* studying the relationship with clinical and epidemiological factors. NHS samples were collected from 185 patients (group C), 136 individuals without leprosy (group Co) who attended the Dona Libânia National Reference Center for Health Dermatology in Fortaleza, Ceará and 121 students from the Christus College in Fortaleza (group GE). Skin biopsies (SB) were collected from 38 individuals in group C. All SN samples were subjected to amplification of the RLEP region using NESTED PCR. To assess the genetic diversity of *M. leprae* between individuals, SN strains from 48 RLEP-positive patients were analyzed using the four variable-number *tandem* repeat *loci* (VNTRs): AC8b, AC9, AC8a and GT9. To study the variability of the strain in the same individual, BP and NS samples from 38 RLEP-positive patients were compared using the fifteen VNTR *loci*: AT17, GGT5, GTA9, AC8b, AC8a, AT15, AC9, 21-3, GAA21, TA18, 6-7, 27-5, TA10, 23-3, 12-5. 69.2% of the cases were RLEP positive, 66.9% in the Co group and 28.1% in the GE group. The fact that the individual was male, belonged to socio-economic class D/E and each year of age he got older increased his chances of being PCR positive by 6.266, 3.083 and 1.046, respectively. In the analysis of the geographical distribution of RLEP-positive individuals, the Co group was the intersection between the C and GE groups and, in relation to the C group, those with positive bacilloscopy and positive RLEP were more clustered than those with positive bacilloscopy and negative RLEP. The AC8b, AC9, AC8a and GTA9 loci showed an allelic diversity considered to be moderately discriminating. Four groups were formed with strains with identical genotypes. The most frequent genotypes were AC8b: 8, AC9: 7, AC8a: 8, GTA9: 10 and AC8b: 7, AC9: 8, AC8a: 9, GTA9: 9. Single genotype strains were detected in younger patients and in patients with a longer time between symptoms and diagnosis. The existence of different strains circulating in the municipality under study and differences in bacillus subpopulations between BP and SN samples from the same individual were found.

Keywords: Leprosy. *Mycobacterium leprae.* Polymerase Chain Reaction.

1 INTRODUCTION

1.1 The genus *Mycobacterium*

The genus *Mycobacterium* belongs to the order Actinomycetales and family Mycobacteriaceae. It consists of aerobic, immobile, non-spore-forming bacilli measuring 0.2 to 0.6 µm wide by 1 to 10 µm long. Even though there are a large number of species of mycobacteria, only a few species cause infections in humans, such as *M. tuberculosis, which* causes tuberculosis, and *M. leprae, which* causes leprosy. The other species are collectively called atypical mycobacteria, causing opportunistic infections, such as the *M. avium* complex, M. *kansassi, M. fortuitum, M. chelonae* and *M. abscessus*. Although the clinical aspects of mycobacteria species vary, most are transmitted via the respiratory tract and have the ability to survive inside macrophages (BARKER, 2006).

The classification of bacteria in the *Mycobacterium* genus is based on their resistance to acids, the presence of mycolic acid containing 60 to 90 carbons which are cleaved by pyrolysis to fatty acid methyl esters of 22 to 26 carbons and the high content (61 to 71%) of guanine and cytosine (G+C) in their genome (COLE et al., 2001; VISSA; BRENNAN, 2001).

Mycobacteria are resistant to many detergents, as well as the usual laboratory stains, due to their lipid-rich cell wall, which makes their surface strongly hydrophobic. Once stained with fuchsin, the bacilli cannot be bleached with acidic solutions or mixtures of alcohol and acid, which is why they are called alcohol-acid-resistant bacilli. The basic structure of the cell wall is composed of an inner cytoplasmic membrane coated with a thick layer of peptidoglycan. The peptidoglycan skeleton is covalently linked to polysaccharides (arabinogalactan) whose ends are esterified to high molecular weight mycolic acids. This layer is coated with polypeptides and a highly antigenic hydrophobic layer of mycolic acids made up of lipids, glycolipids and free glycolipopeptides. These lipids make up 60% of the dry weight of the cell wall. The peptide chains are also considered biologically important antigens, stimulating the host's cellular immune response (SCOLLARD et al., 2006; VISSA; BRENNAN, 2001). The cell wall is also responsible for other properties of these bacteria, such as resistance to common antibacterials and the ability to aggregate (BARKER, 2006; COLE et al., 2001).

1.2 *Mycobacterium leprae*

M. leprae is an intracellular, immobile, non-spore-forming bacterium that appears as a straight or slightly curved bacillus, measuring 1.5 to 8 µm long by 0.2 to 0.5 µm wide (MONOT et al. 2005; WALKER; LOCKWOOD, 2006). Its peptide chains constitute biologically important antigens that stimulate the patient's cellular immune response to infection, such as the PGL-I antigen (phenolic glycolipid I) which can be used in immunodiagnosis (BARKER, 2006).

Inoculation of *M. leprae* was first achieved by Shepard in 1960 in a mouse paw, obtaining a localized lesion. In 1971, Kirchheimer and Storrs succeeded in infecting armadillos, verifying the involvement of the skin, peripheral nerves, bone marrow, liver, spleen, lymph nodes, lungs, meninges and eyes (SHEPARD, 1960; KIRCHHEIMER and STORRS, 1971). *M. leprae* is the only species of the genus that cannot be cultivated "*in vitro*" and is therefore only cultivated in animal models such as armadillos and mouse feet (REES, 1988; SHEPARD, 1960; TRUMAN et al., 1986).

The optimum temperature for its reproduction is between 27°C and 33°C, which explains its predilection for colder areas of the body, such as the skin, peripheral nerves and upper airway mucosa. In the skin, *M. leprae* is preferentially found in macrophages, while in neural tissue, it is located in *Schwann* cells, where it can remain for a long period of time without stimulating an immune response. In skin smears and histopathological sections, the bacilli are seen isolated, in varied groupings or in arrangements called globia, peculiar to *M. leprae,* which result from the union of bacilli through a substance called glia (GOULART; GOULART, 2009).

This mycobacterium has a low frequency of mutations and a generation time of 24 hours, with a slow generation time of approximately 11 to 13 days in the logarithmic phase in the plantar pad of an immunocompetent mouse, which limits the genetic diversity of the species (BRITTON; LOCKWOOD, 2004; SASAKI et al., 2001). It is suggested that the slow growth of *M. leprae* is due to its categories of functional genes being considerably smaller than in other mycobacteria, with some almost disappearing completely. Although the main anabolic pathways are relatively intact, there has been an extensive reduction in genes required for lipolysis. This is the main catabolic pathway used by intracellular mycobacteria to obtain their energy from the degradation of host-derived lipids and fatty acids. Likewise, energy metabolism pathways are minimal, such as the use of common carbon sources, such as acetate and galactose, and ATPs that cannot be generated from NADH oxidation. In addition, the anaerobiosis and electron transfer systems, along with the biosynthesis and transport systems needed to produce the prosthetic groups, have been lost. This implies that *M. leprae* is restricted to growth on little carbon source and catabolism is severely limited (COLE et al., 2001).

1.3 The *M. leprae* genome

From strains of *M. leprae* derived from armadillos, it was possible to extract enough DNA to sequence the complete genome of the TN (Tamil Nadu) strain, originally isolated from a patient in Tamil Nadu, India. Thus, it was shown that the *M. leprae* genome contains 3,268,210 bp, 57.8% G+C content, 49.5% (1293) pseudogenes, 2% repetitive sequences, of which can be dispersed (RLEP, REPLEP, LEPREP and LEPRPT) or in *tandem* (VNTR) (CLARK-CURTISS; WALSH, 1989; COLE et al., 2001; WOODS; COLE, 1990).

While *M. tuberculosis* has a genome of 4,411,532 bp, which can encode 3,959 proteins, *M. leprae* encodes only 1,604 proteins. This drastic reduction in the number of genes in *M. leprae* compared to *M. tuberculosis* and the high number of pseudogenes, affects aspects of its metabolism, justifying the impossibility of this microorganism to grow "*in vitro*". This minimalist evolutionary process has made M. *leprae* completely adapted to the human organism (SINGH; COLE, 2011; VISSA; BRENNAN, 2001).

Mutations are found in genes involved in regulation (transcriptional regulators, repressors and activators), detoxification (peroxidase encoders), DNA repair (*mutT*, *dnaQ*, *alkA*, *dinX* and *dinP genes*) and in the transport or efflux of metabolites such as amino acids (arginine, ornithine, D-alanine, D-serine and glycine), peptides, cations (magnesium, nickel, mercury, ammonia, ferric and ferrous ions and potassium) and anions (sulphate and phosphate). However, the genes for the synthesis of most of the smaller molecules such as amino acids, purines, pyrimidines and fatty acids, and for the synthesis of macromolecules such as ribosomes, RNA and structural proteins are preserved in the genome (VISSA; BRENNAN, 2001).

In terms of genetic reduction, *there are* fewer genes in almost all categories, but the most notably affected are the insertion sequences and the families of glycine-rich proteins that have glutamic acid-proline or glutamic acid-proline-proline at the amino terminus. These proteins can confer antigenic variation. Thus, while genes are required for transmission, establishment and survival in the host, the genetic reduction of *M. leprae* can be compensated for through host-dependent parasitic life (VISSA; BRENNAN, 2001).

The mechanism by which pseudogenes arise is not yet known, but different explanations have been proposed, such as an altered or deficient response to environmental or host-mediated stresses that lead to DNA damage. The damage may be due to the loss of sigma factors and the lack of activity of DnaQ-mediated DNA polymerase III (which polymerizes the epsilon subunit of the DNA polymerase III holoenzyme, whose function is to remove incorrectly incorporated DNA bases), contributing to the accumulation of mutations in genes (BABU, 2003; LIU et al., 2009; TYAGI; SAINI, 2004). An important characteristic of *M. leprae* pseudogenes is the presence of a large number (ranging from 1 to 40) of stop codons. The minority (25%) of transcribed pseudogenes have start codons, while the majority (67%) contain more than five stop codons (NAKAMURA et al., 2009). Pseudogenes tend to be eliminated from the genome, but it is intriguing that *M. leprae* has accumulated and maintained many of them (SINGH; COLE, 2011).

Repetitive sequences are common constituents of the genome of all living organisms, however, they occur more frequently in higher eukaryotes. There are two main forms of repetitive DNA in the bacterial genome; dispersed sequences and *tandem* sequences that align repeatedly (SINGH; COLE,

2011).

The dispersed repetitive sequences may correspond to duplicated genes or mobile genetic elements present in numerous copies, such as insertion sequences (IS). The small size of the *M. leprae* genome was mainly due to recombination events between these repetitive sequences. In *M. leprae* there are only trace elements of insertion sequences, more than 26 transposase gene fragments and four families of dispersed repeats: RLEP (37 copies); REPLEP (15 copies); LEPREP (eight copies); and LEPRPT (five copies) (COLE et al., 2001; MONOT et al., 2009). RLEP is among the genome sequences that do not encode regions. The presence of RLEP at the 3' end of genes and often within pseudogenes suggests that these sequences are the remnants of transposons that have already lost their ability to undergo transposition.

Previous studies on RLEP based on restriction fragment length polymorphisms (RFLPs) did not detect diversity, as was later confirmed by genomic comparison of *M. leprae* strains (COLE et al., 2001; WILLIAMS et al., 1990).

Tandem repeats can be relatively simple, such as multiple repeats of two or three nucleotide sequences, or more complex, such as *tandem* duplications of large chromosomal segments. An important form of *M. leprae* DNA polymorphism results from variations in *tandemly* repeated DNA sequences (VNTRs).

M. leprae exhibits very little genetic variability, however, single nucleotide polymorphisms (SNPs) can occur in some regions of its genome. SNPs can be restricted to a particular strain or shared by strains. The first SNP typing system was based on three informative SNPs and used to type approximately 400 different strains from 28 countries around the world. This approach revealed only four combinations (SNP types 1, 2, 3 and 4), with geographic associations, providing evidence for a global evolutionary scheme of *M. leprae*. Most samples from Central Asia presented type one, while type two predominated in Ethiopia; type three in Europe, North Africa and the Americas and type four was most frequently found in West Africa and the Caribbean. The authors suggest that leprosy originated in East Africa, as this region has type two, which is rarer and probably the oldest genotype of *M. leprae* (MONOT et al., 2005).

Subsequently, 84 markers (78 SNPs and six insertion and deletion sequences) were found, thus allowing the classification of *M. leprae* strains into 16 different genotypes, subtypes of the four SNPs (1A-4P), once again showing a phylogeographic relationship (MONOT et al., 2009). In a country with endemic leprosy cases, this system can help uncover possible sources of infection through immigration or the patients' history of residence abroad. This system is stable and reliable for the classification of strains in wider regions, but for short-range transmission studies it is best to use genotyping through VNTRs.

1.4 *M. leprae* genotyping

In recent years, methodologies for genotyping *M. leprae* isolates have complemented the conventional epidemiology of infectious diseases. The selection of polymorphic genomic markers became feasible for genotyping strains after the publication in 2001 of the complete genome sequence of *M. leprae* strain TN (COLE et al., 2001).

The first markers in the *M. leprae* genome to show polymorphism were STRs (*short tandem repeats*). Almost all STRs are within intergenic regions or pseudogenes and are unlikely to interfere with biological pathways. Few STRs, particularly minisatellites, are inside genes. STR loci, classified as microsatellites (repeat units smaller than 6 bp) or minisatellites (repeat units from 6 to 100 bp), are excellent sources of polymorphism.

STRs have a higher mutation frequency due to *strand mispairing* during DNA replication, leading to clonal instability. *Slipped* strand *mispairing* is an error caused by the DNA polymerase "slipping" from the template strand during replication, which ends up failing to incorporate STRs or adding new ones, forming a new strand of DNA with a different number of repeats from the template strand. This hypothesis is the most widely accepted to explain the appearance of small variations in the number of repeats between the different alleles within the locus, and is most common in microsatellites. Thus, while this dynamism provides greater discriminatory power, the lower stability and reproducibility of some of these loci needs to be considered and they should therefore be carefully selected in order to extract meaningful epidemiological information and/or conclusions (LOVETT, 2004; MONOT et al., 2008; REYES; TANAKA, 2010).

Whole genome sequencing of the Br4923 (Brazil), Thai53 (Thailand) and NHDP63 (*National Hansen's Disease Program-USA*) strains showed that the Br4923 genome comprises 3,268,071 bp, which compared to the TN genome was reduced by 141 bp due to the higher number of copies of *Variable Number T*andem Repeat (VNTR) loci in the TN strain. In total, the Br4923 and TN genomes differ in 185 loci, which include 146 SNPs, 31 VNTR regions, and eight insertion-deletion events (Indel). Comparison of the genome of strains Thai53 (Thailand) and NHDP63 (USA) with strains TN and Br4923 revealed only 215 polymorphic sites and more than 99.995% of the genome identical between these geographically distant strains and some recombination events between dispersed repeats (RLEP) (MONOT et al., 2009). Almost 25% of all SNPs occur in these dispersed repeat sequences, which occupy only 1.16% of the genome. This may indicate that recombination events between different copies of repetitive elements result in the dispersion of a specific SNP (SINGH; COLE, 2011).

STRs have been successfully used for typing *M. tuberculosis as* well as other mycobacteria (ALLIX-BÉGUEC et al., 2008; SUPPLY et al., 2006; THIBAULT et al., 2008). In *M. leprae,* the first STR reported as a genetic marker was a six-bp intragenic sequence in the *rpoT* gene (MATSUOKA; KASHIWABARA; NAMISATO, 2000), which presented itself as two alleles within the *M. leprae* gene; the strains from patients in Japan (with the exception of Okinawa) and Korea showed three copies of the repeat unit, while those from Brazil, Haiti, and Okinawa, Japan, showed four copies. The second was a repeat element from a pseudogene (TTC) (SHIN et al., 2000), of which alleles ranging from 10 to 37 copies were detected. These sequences showed varying numbers of *tandem* repeats (VNTRs) when sequenced in different isolates.

These studies inspired other researchers to initiate a more comprehensive analysis of the potential of other STRs in *M. leprae*, because when several independent STR loci from clinical samples are evaluated, unique patterns or "*fingerprints*" emerge. Thus, around 50 VNTR loci have been identified in *M. leprae* (GILLIS et al., 2009; GROATHOUSE et al., 2004; MATSUOKA; KASHIWABARA; NAMISATO, 2000; MATSUOKA et al., 2004a; SHIN et al., 2000; TRUMAN et al., 2004; YOUNG et al., 2004; ZHANG; BUDIAWAN; MATSUOKA, 2005).

In 2004 Groathouse et al. (2004) pre-selected 44 loci (including *rpoT* and TTC) from the *M. leprae* genome and screened 11 STR loci, of which 9 were polymorphic from four human isolates and tested by passaging them in armadillos. Five were minisatellites (6 to 50 bp repeat units) and four were microsatellites (1 to 5 bp repeat units). These authors detected the VNTRs between the samples by the difference in the migration of the PCR product in the 3% agarose gel electrophoresis, demonstrating that there were at least two alleles for the loci tested. They also sequenced some PCR products to confirm that the alleles at the loci vary in size due to the VNTRs.

Matsuoka et al. (2004) applied the microsatellite locus (TTC) 21 to genotype *M. leprae* obtained from nasal swabs and skin smear slides of patients from a village or a single household in Indonesia. While Young et al. (2004) combined the VNTR loci (AT) 15, (GTA) 9, and (TTC) 21 for the identification of short- and long-range transmission chains of *M. leprae* in areas in and around the city of Hyderabad, India. Truman et al. (2004) demonstrated that the (TTC) 21, AT17, TA18 and (GTA)9 loci were stable after multiple passages in armadillos and mouse paws, while Matsuoka et al. (2004) proved the stability of the (TTC) 21 locus after eleven passages in mouse paws.

Monot et al. (2008) mapped five STR loci of *M. leprae* in patients from Mali, Africa and concluded that in this area of endemicity, multiple dominant strains are found and that additional genetic markers are needed for genotyping.

In studies in China and the Philippines, *M. leprae* was genotyped in skin biopsy samples and BAAR slides before the start of treatment and two months apart during treatment. These studies revealed intra-patient variability (mainly in di- or trinucleotide repeats such as AT17 and GAA21), while other loci remained stable, thus allowing the authors to establish transmission relationships between isolates from families or the same city (SAKAMURI et al., 2009; XING et al., 2009).

Genetic variability between strains of *M. leprae* in the family context was observed by Sakamuri et al (2009). In this study, the authors studied three patients from the same family, two of whom had corresponding strains, but one of the patients differed in four loci of 13 VNTRs. When epidemiological information was taken into account, it was found that this family had a history of five other leprosy patients (SAKAMURI et al., 2009).

Using a reference strain, assays were carried out to test the sensitivity, specificity and reproducibility of the 16 loci: (AC)8a, (AC)8b, (AC)9, (AT)15, (AT)17, (TA)10, (TA)18, (GGT)5, (GTA)9, 6-7, 12-5, 18-8, 21-3, 23-3, (GAA) 21 and 27-5 (GILLIS et al, 2009). As a measure of lower limit sensitivity, the *M. leprae* detection capacity of each primer pair was analyzed using PCR reactions at concentrations of 1000 *M. leprae* / PCR (200,000,000 bacilli/ml), 100 M. *leprae* / PCR (200,000 bacilli/ml) and 10 *M. leprae* / PCR (2,000 bacilli/ml). Five identical reactions were carried out for each concentration, and the products were analyzed on an agarose gel, observing the intensity of the band on the gel. As a result, from a concentration of 10 *M. leprae* / PCR all the primer pairs produced a visible fragment on the gel and only a single dominant band, except for locus (GAA) 21 which produced a weak secondary band. At high concentrations of *M. leprae*, the loci (AC) 8a, (AT) 15, (AT) 17, (TA) 10, (TA) 18, 21 (GAA), 18-8 and 6-7 produced at least one other minor band visible on the gel. To assess specificity and reproducibility, the product of the PCR reactions at a concentration of 1000 *M. leprae* / PCR were sequenced by two different sequencers. All loci were identified as specific to *M. leprae* by both sequencers using either forward or reverse primers. To assess the stability of 13 loci during infection, NHDP60 strains of *M. leprae* were inoculated into armadillos, representing four cycles of infection over a period of five years. All loci except (GAA)21 were found to be stable over time (GILLIS et al., 2009). Thus, all the primer pairs passed the sensitivity, specificity and reproducibility criteria.

Variability should also be assessed in terms of whether changes in one or more STRs allow the detection of clusters of the bacillus in a given region.

Kimura et al (2009) reported on the development, testing and reproducibility of the *M. leprae* multiple VNTR loci analysis (MLVA) technique on strains derived from armadillos and clinical materials, as well as variations in the number of repeats of the proposed loci. These authors standardized the amplification of 17 loci in four multiplex PCR combinations. All the combinations were optimized with DNA from the so-called NHDP63 strain obtained from the tissue of an armadillo infected with bacilli from a leprosy patient, and this DNA was then used as a reference for VNTR mapping.

To test the feasibility of multiplex PCR and FLA (*Fragment Length Analysis*) on clinical samples, Kimura et al (2009) used DNA from 30 stored biopsy samples from leprosy patients. In addition, a set of samples was subjected to multiplex PCR using unlabeled primers and DNA sequencing. For each combination, at least three clinical DNA isolates were randomly tested. Thus, it was found that the alleles determined by sequencing the amplified ones were in complete agreement with those estimated by FLA.

To test the reproducibility of FLA, Kimura et al (2009) also performed multiplex PCR several times using DNA from the NHDP63 strain and the products of each PCR were subjected to FLA on separate days. Clinical samples were also subjected to multiplex PCR and FLA with *Qiagen* multiplex enzyme and another enzyme (*AmpliTaq Gold*; *Applied Biosystems*). In general, the sizes of the peaks and patterns remained stable between the independent PCR and FLA experiments, and it was also shown that the variation due to DNA polymerase is minimal and insignificant. Furthermore, when the multiplex PCR products were stored for FLA, there were no significant changes in the profiles, indicating that the products can be stored for up to a year at -20 °C.

Also in the study by Kimura et al (2009), Nei's (1978) genetic diversity was 0.5463 (for 16 loci) in 21 armadillo samples and the TN strain and 0.54 (for 15 loci) in 30 clinical isolates from the Philippines. For these same loci, the genetic diversity of 27 isolates, mainly of Japanese origin, reported by Zhang et al. (2005), was 0.6. These diversity indices are comparable to VNTR typing systems for other bacteria. With *Bacillus anthracis*, an eight VNTR-locus system was used for typing, showing an average diversity of 0.52 for 89 isolates (KEIM et al., 2000). With *Salmonella enterica* serovar *S. Enteritidis*, Malorny et al. (2008) demonstrated the discrimination of isolates in a typing system based on nine VNTR loci, even with individual allelic diversity ranging from 0.07 to 0.65 (MALORNY; JUNKER; HELMUTH, 2008). For typing *M. tuberculosis* strains, Supply et al. (2006) demonstrated an average genetic diversity of 0.58 for 24 loci (SUPPLY et al., 2006).

In this way, MLVA (Multi-Locus Analysis of VNTR through FLA) was later applied in population-based studies in Cebu, Philippines (SAKAMURI et al., 2009), and it was established that the ability to distinguish strains with only 15 loci and the fact that it allows PCR *stutter* products to be distinguished from the main peak of the allele, demonstrates the potential of MLVA for molecular typing of *M. leprae*. Thus, microsatellites tend to show greater allelic diversity than minisatellites and when these are combined, they are considered useful in differentiating samples from different countries and geographical regions (SAKAMURI et al., 2009). It is therefore suitable for monitoring the spread of *M. leprae* over short distances, such as within a region or a city.

1.5 Leprosy

The earliest written records describing the disease come from ancient Indian texts in the Sushruta

Samhita around 600 BC, which were certainly passed down orally over previous centuries. There are also descriptions of the disease in ancient Egyptian texts. The oldest skeleton with leprosy found was at Balathal in India, 4,000 years old (LOWE, 1947; HULSE, 1972).

The main gateway to Europe was through Greek traders, colonizers and soldiers between the 10th and 15th centuries. During the 17th and 18th centuries, the disease declined sharply on the European continent as a result of better socio-economic conditions. During the period of colonization, Latin America became the great endemic focus of leprosy, mainly by Portuguese colonizers and African slaves. At the same time, Scandinavian colonizers spread leprosy in North America (LOWE, 1947; MONOT et al., 2005).

The disease entered Brazil from various points along the coast, with the first Portuguese settlers, mainly Azoreans, and the spread was contributed to by African slaves. The first recorded cases were in Rio de Janeiro in 1600. After that, other outbreaks of the disease were identified, mainly in Bahia and Parà (ANTONIO; SANTOS; ARMAUER, 2008).

The cause of the disease was only known after 1873, when the Norwegian Gerhard Henrik Armauer Hansen, when examining the leprous cells found in the nodules of a patient, noticed the presence of the bacillus under the microscope (GOULART; GOULART, 2009).

Until the 1950s, treatment was basically with chaulmoogra oil, a herbal medicine from India that did not act directly on the bacillus, but only on the effects of the disease that had already taken hold in patients. Chaulmoogra and isolation were considered the best ways to treat leprosy. Chemotherapy drugs were only used after the Second World War, with advances in the chemical-pharmaceutical industry, laboratory tests and the discovery of the etiological agents of various diseases (MACIEL et al., 2003).

In Brazil, the policy of controlling the disease, after the 1930s, was mainly carried out by building leprosaria in the units of the federation, some of which even had more than one hospital (LIMA et al., 2003).

In 1958, the Seventh International Congress of Leprology was held in Tokyo, confirming that the form of contagion was not hereditary and that there was a possibility of a cure with antibiotics. Thus, isolation, which had been questioned since the 1920s as a factor in the cure, should no longer be recommended as a fundamental element of treatment for the disease (MACIEL et al., 2003).

In Brazil, the approval of Decree No. 165 of May 14, 1976, which officially changed the name of the disease to leprosy, resulted in widespread mobilization to reduce the prejudice surrounding the word leprosy and its derivatives (MACIEL et al., 2003).

In June 1981, the creation of MORHAN (Movement for the Reintegration of People Affected by

Leprosy) was a milestone in ensuring the rights of patients, guaranteeing the social reintegration of those affected by leprosy and discussing the deactivation of leprosaria, which had their role redefined (MACIEL et al., 2003).

The World Health Organization (WHO) declared that leprosy was eliminated as a global problem in 2000, despite the notification of 719,330 new cases that same year. Of the 122 countries considered endemic in 1985, 107 reached the target set by the WHO (BRITTON; LOCKWOOD, 2004).

1.6 Epidemiological aspects

Polychemotherapy (MDT) was instituted by the global leprosy control programs, which has cured more than 14 million people in the last 20 years. Although this correlates with more than a 90% reduction in the prevalence rate, the chain of transmission has not been broken with around 250,000 new cases of leprosy still being registered annually and concentrated in a few countries. This translates into the identification of a new case approximately every 2 minutes, highlighting the need to continue with the commitment to disease control and leprosy research. The regions most affected by leprosy in 2012 were: 25 countries in Africa, 28 in the Americas, 20 in the Mediterranean, 11 in Southeast Asia and 31 in the Western Pacific. According to WHO (WHO, 2013) and based on reports from 130 countries, there was an increase in the recorded global prevalence and detection of new cases in 2012, which was 181,941 cases (0.34), compared to 189,018 (0.33) at the end of the first quarter of 2013 (TALHARI et al., 2012).

The distribution of new cases detected in 2012 by the WHO was similar to the findings of previous years. Southeast Asia accounted for 71% of new cases detected worldwide, the Americas for 16%, Africa for 9%, the Eastern Mediterranean for 2% and the Western Pacific for 2% (WHO, 2013).

In Brazil, in 2012, the leprosy prevalence coefficient was 1.51 cases/10,000 inhabitants, which represented a 12% reduction compared to the coefficient in 2004 (1.71 cases/10,000 inhabitants). In the same year, 33,303 new cases were diagnosed, of which 2,246 (7%) were in children under 15, giving an overall detection rate of 17.2/100,000 inhabitants, which is considered very high (MINISTÉRIO DA SAÙDE, 2013).

The states of Mato Grosso, Maranhao and Tocantins had the highest prevalence rates (between 5 and 9.99 cases/10,000 inhabitants) compared to the states of the South, Minas Gerais, Rio de Janeiro and Sao Paulo in the Southeast, plus Rio Grande do Norte in the Northeast (MINISTÉRIO DA SAÙDE, 2013).

In 2012, the state of Cearà reported 2,136 new cases, with an overall detection rate of 24.82 and a rate of 5.3/100,000 inhabitants in children under 15, demonstrating the existence of active leprosy transmission centers, with an annual average of 5.9% of the total reported cases of the disease in this age group. Among the 184 municipalities in Ceará, 29 (16%) did not report any cases in 2010, 32 are considered hyperendemic, of which 14 had less than 10 new cases. The capital Fortaleza had 24.4 cases/100,000 inhabitants in 2012. However, there has been a significant drop in the overall detection rate of 2.2 cases/100,000 inhabitants per year over the last 10 years (SECRETARIA DA SAÙDE DO GOVERNO DO ESTADO DO CEARA, 2013).

The manifestation of the disease in the bacillus-infected person and its different clinical forms depend, among other factors, on the parasite/host relationship and can occur after a long incubation period, which varies from two to seven years. In addition to individual conditions, other factors related to endemic levels and unfavorable socioeconomic conditions, as well as precarious living and health conditions and the high number of people living in the same environment, influence the risk of falling ill (JOYCE, 2012; KERR-PONTES et al., 2004).

Leprosy can affect people of all ages, of both sexes, but it rarely occurs in children. It has been observed that children under the age of fifteen fall ill more often when the disease is more endemic. In general, the highest case detection rates are in males, and the risk of exposure is generally considered to be the responsible factor (BRITTON; LOCKWOOD, 2004; JOYCE, 2012).

Man is considered the only natural reservoir of *M. leprae*, although there have been reports of naturally infected wild animals such as armadillos and monkeys (MEYERS; GORMUS; WALSH, 1992; TRUMAN et al., 1986). The upper airways are the main route of entry and elimination of *M. leprae*. However, the precise mechanism of transmission of *M. leprae* is still unknown (BRITTON; LOCKWOOD, 2004; JOYCE, 2012).

M. leprae is capable of surviving outside the human body for several months under unfavorable conditions. In addition, it is believed that environmental reservoirs of *M. leprae* cannot be excluded due to the existence of a considerable number of epidemiological and microbiological observations which indicate that environmental sources may play an important role in the transmission of the disease. Various environmental sources have been studied in order to prove indirect contamination and include soil, vegetation, water, arthropods, armadillos and monkeys (BRITTON; LOCKWOOD, 2004; LAVANIA et al., 2008; SASAKI et al., 2001).

1.7 Classifications and clinical manifestations

Leprosy has been classified in various ways. In 1937, Rabello described the indeterminate form (I) and the tuberculoid (TT) and virchowian (V) polar forms (RABELLO, 1937). At the International Congress of Leprology held in Madrid in 1953, the criteria proposed by Rabello were maintained, with the addition of a new group of patients, which was called "borderline" or dimorphic (R et al., 2008).

In the 1960s, Ridley and Jopling proposed a modification to the Madrid classification, introducing the concept of the spectral classification of leprosy, subdividing the dimorphs into dimorph-tuberculoid (DT), dimorph-dimorph (DD) and dimorph-virchowian (DV). The concept of disease polarity was maintained and this classification, called spectral, was based on clinical and histopathological parameters (RIDLEY and JOPLING, 1966). In the studies by Ridley and Jopling (1966), only cases with skin lesions, associated or not with nerve lesions, were included. The pure neural form was not included in the spectrum, nor was the indeterminate group of the Madrid Classification. Regarding these two presentations of the disease, Ridley and Jopling state that in the case of pure neural leprosy, nerve thickening appears early in TT cases, but can be seen in all cases, with the exception of V cases. It is reported that TT patients have one or two nerves affected, while DT or DD patients have several nerves affected (RIDLEY and JOPLING, 1966).

With a view to treating patients with multidrug therapy, in 1982 the WHO adopted the paucibacillary form (PB) for patients I, TT and most DT, all of whom had a baciloscopic index (BI) of less than 2+ on the Ridley Scale, and the multibacillary form (MB) for patients DD, DV and V, all of whom had a BI equal to or greater than 2+. However, in recent years, the WHO has been dividing patients into the paucibacillary form and the multibacillary form, which is an operational classification aimed at treatment and which takes into account the number of skin lesions (RIDLEY, 1974; WHO, 2013; PARDILLO et al., 2007).

In Brazil, the Ministry of Health classified leprosy according to Ordinance 817 of July 26, 2000, considering paucibacillary patients to be those with less than five skin lesions and/or with involvement of one nerve trunk and multibacillary patients to be those with five or more skin lesions and/or with involvement of more than one nerve trunk. The multibacillary group includes all cases with positive bacilloscopy, whatever the BI (MINISTÉRIO DA SAÙDE, 2013).

It is accepted that leprosy V can evolve from the indeterminate form or present as such from the outset. Its chronic evolution is characterized by symmetrical, slightly erythematous, poorly delimited macules, which progress to a generalized and diffuse infiltration of the skin, upper airway mucosa, eyes, testicles, nerves, and may also affect the lymph nodes, liver and spleen. Pules and nodules can be seen on the skin. Infiltration is diffuse and more pronounced in areas with lower body temperature, such as the face and limbs. Hά hair thinning on the limbs, eyelashes and eyelids. Hair loss in these areas is called madarosis. Infiltrations of the face, including the auricles, with madarosis and hair loss form a condition known as *facies* leonina. The maintenance of the hair is due to the scalp being a site of higher temperature. Nerve involvement occurs in the skin muscles, vascular innervation and nerve trunks. The latter will present functional deficiencies and late sequelae (LOCKWOOD; SUNEETHA, 2005; WALKER; LOCKWOOD, 2006).

The macrophages of Virchowian (V) patients are unable to eliminate the phagocytosed bacilli, which allows them to survive and multiply inside these phagocytes, which then turn into Virchow cells, full of bacilli (with large amounts of lipids in their wall) and fat droplets. These patients have abundant bacilli on their skin and nasal mucosa. However, V patients are not immunocompromised hosts. The immunological allergy associated with V patients is specific to *M. leprae* antigens (BARKER, 2006; FOSS; CAROLINA; MOTTA, 2012).

V patients have a predominantly humoral response, represent the multibacillary form and in untreated cases are an important focus of infection and reservoir of the disease. The V form has a strongly positive bacilloscopy (FOSS; MOTTA, 2012).

In TT leprosy, the lesions are well-defined, few in number, hypopigmented, anesthetic and asymmetrically distributed. There are plaque lesions with papular borders and erythematous areas of skin, whose slow growth leads to atrophy inside the lesion with scaling of the borders. There is also the infantile variety, which is mainly located on the face and can manifest as nodules or plaques (SCOLLARD et al., 2006; WALKER; LOCKWOOD, 2006).

TT patients have a Th1-type cellular immune response, which restricts the growth of the pathogen, constituting the paucibacillary form of leprosy. Despite the possibility of a spontaneous cure, the advice is that cases should be treated to reduce the time the disease takes to develop and the risk of neural damage (FOSS; MOTTA, 2012; LOCKWOOD; SUNEETHA, 2005).

The indeterminate group is not part of Ridley and Jopling's spectral classification. It includes patients who have the initial manifestations of leprosy and is a benign form of the disease. It is considered a transitional form, with an as yet undefined immunological pattern. In indeterminate leprosy, the lesions appear after an incubation period of between two and five years. There are hypochromic spots with altered sensitivity which can occur in small numbers and on any area of the skin. Generally, only thermal sensitivity is altered, with no involvement of nerve trunks. Histopathology shows no evidence of granuloma and the inflammatory infiltrate is nonspecific. Sputum smear microscopy is negative. This form, which is considered to be the first clinical manifestation of leprosy, can progress to a cure or to another clinical form (R et al., 2008; WALKER; LOCKWOOD, 2006).

The dimorphic group is made up of patients who fall into the interpolar area, as they cannot be classified into one of the polar types of leprosy. These patients have lesions with structures from both lepromatous and tuberculoid leprosy in reaction. Dimorphic leprosy is characterized by its immunological instability, which means that there is great variation in its clinical manifestations, whether on the skin, nerves or systemic involvement. The skin lesions are numerous and their morphology mixes aspects of DV and DT, and there may be a predominance of one or the other

(ARMOUR; SCOLYER; BARNETSON, 2005; FOSS; MOTTA, 2012).

The DT form is similar to tuberculoid, but the lesions are more numerous. The DV form differs from the virchowian form in that the lesions are more defined and have areas of healthy skin between them. The DD form is an unstable form that can vary from DT to DV. It is characterized by the presence of plaques with raised, well-defined borders, with islands of apparently healthy skin. Sputum smear microscopy can be negative or positive with a variable bacillary index (ARMOUR; SCOLYER; BARNETSON, 2005; R et al., 2008).

In the pure neural form, there are no skin lesions. Há involvement of one or more peripheral nerves with thickening of the nerve trunk and early nerve damage, especially when it affects sensory-motor nerves (SCOLLARD et al., 2006; WALKER; LOCKWOOD, 2006). In Ridley and Jopling's spectral classification, this form is also not included, although it is described as a subtype of the tuberculoid and dimorphic forms (RIDLEY; JOPLING, 1966).

Reactive outbreaks represent inflammatory episodes that are interspersed in the chronic course of leprosy. The most important types of reaction are the reverse reaction or type 1 reaction and the type 2 reaction or erythema nodosum of leprosy (CUEVAS et al., 2007; FOSS; MOTTA, 2012).

Type 1 reaction is a hypersensitivity reaction that can occur in 30% of patients with dimorphic forms. It is characterized by the exacerbation of pre-existing lesions which become swollen, erythematous and shiny, and may even ulcerate. New lesions may appear and neuritis is frequent and may be the only clinical manifestation (CUEVAS et al., 2007; FOSS; MOTTA, 2012).

The type 2 reaction can affect 20% of Virchowian patients and 10% of patients with the VD form, and is more frequent the greater the skin infiltration and bacterial index. It is triggered by deposits of immune complexes and is associated with high concentrations of tumor necrosis factor-α (TNFa). Erythema nodosum is the typical epidermal lesion and is characterized by painful erythematous lesions of varying sizes, including papules and nodules located on any region of the skin. The presence of a high fever can accompany this reaction (FOSS; MOTTA, 2012; RADA et al., 2012).

1.8 Treatment

Leprosy treatment includes intensive chemotherapy, suppression of relapses, prevention of physical disabilities, physical and psychosocial rehabilitation. The therapeutic regimen depends on the operational classification of the case. The drugs used are rifampicin, dapsone and clofazimine. The drug is supplied free of charge in Brazil. There are also alternative treatments using rifampicin, ofloxacin and minocycline to treat single skin lesions in paucibacillary patients (WHO, 2013).

1.9 BCG vaccine

The BCG (Bacillus Calmette Guèrin) vaccine prevents mycobacterial diseases, including leprosy and tuberculosis. In neonates, the BCG vaccine is routinely used in most parts of the world, including Brazil, to prevent tuberculosis (RODRIGUES et al., 2007).

Studies carried out in India and Brazil have shown that BCG protects against leprosy. In the Brazilian study by Rodrigues et al (2007), the results suggest that vaccinating neonates with BCG can have an important impact on the transmission of leprosy and that even environmental mycobacteria do not interfere with the effectiveness of the vaccine, at least in the Amazon region (RODRIGUES et al., 2007). The Indian study showed a vaccine efficacy of 54%, with the greatest protective effect seen in the multibacillary form (68%) and suggested that the vaccine is very effective in preventing the transmission of *M. leprae* within a community (CUNHA et al., 2004; ZODPEY; AMBADEKAR; THAKUR, 2005).

In another study in Brazil, the results showed a high efficacy of 86% of BCG vaccination against leprosy up to the age of 29 (RODRIGUES et al., 2007). This estimate of protection is within the range provided in three other older studies carried out in different states of Brazil, in which a protection of 74% was found in Manaus, in the North (CUNHA et al., 2004), 81% in the central region (RODRIGUES et al., 1992) and 90% in Sao Paulo, in the Southeast of the country (LOMBARDI et al., 1996).

Rodrigues et al. (2007) found that the protective effect of BCG against leprosy can last at least 30 years, and suggest that it could eventually last more than 40 years. Their results also showed that adults with a BCG scar around the age of 40 had a lower risk of contracting leprosy. Thus, this study adds to the new and growing evidence that BCG's protection against tuberculosis and leprosy may last longer than previously believed (RODRIGUES et al., 2007).

1.10 Diagnosis

The most widely used diagnosis of leprosy is based on clinical findings and confirmed by bacilloscopy. Other tests may be necessary, such as histopathological examination of the skin, nerve biopsy, the Mitsuda reaction and serological tests (SASAKI et al., 2001).

Histopathological examination of the skin is important for reducing doubts in the diagnosis, as well as for classifying the disease. Histologically, the presence of neural inflammation differentiates leprosy from other granulomatous disorders (BRITTON; LOCKWOOD, 2004).

Sputum smear microscopy is the most useful complementary test in diagnosis, being easy to perform and low cost. Its high specificity makes it possible to identify patients with the greatest potential for infection and those with the greatest risk of recurrence. It is collected from the earlobes, elbows and suspicious lesions. It is stained using the Ziehl-Neelsen method and the result is presented as a baciloscopic index (MINISTÉRIO DA SAÙDE, 2013).

Mitsudin, the name adopted in Brazil for lepromine, is still widely used in the northern region of Brazil. It consists of an intradermal injection of heat-inactivated Hansen's bacilli, mechanically extracted from the lepromas of patients who have not been treated or from the tissues of armadillos infected with *M. leprae*. The reaction is considered positive with the appearance of a weak infiltration, papule or nodule more than three millimeters in diameter. This is due to the bacilli in this suspension being phagocytized by macrophages, which then turn into epithelioid cells. Thus, the Mitsuda reaction is negative in V patients, positive in TT patients, weakly positive in dimorphic patients and variable in the group with the indeterminate form. In other words, a positive reaction only indicates that an individual's macrophages are capable of destroying both dead and live Hansen's bacilli, and is not very useful in diagnosing the disease. This test is used to classify the disease and define the prognosis, so it is not used for diagnosis (BEIGUELMAN, 2002; SCOLLARD et al., 2006).

The application of the serological test for the detection of anti-PGL-I antibodies is yet another diagnostic resource in leprosy, especially in the multibacillary forms. The most precise indications would be for the diagnosis of subclinical infection, sero-epidemiological mapping, therapeutic follow-up and early detection of disease relapse. However, the test is not useful for diagnostic purposes in an endemic community (FROTA et al., 2010a). Studies have documented the relationship between leprosy endemicity levels in the population and seroprevalence in different parts of the world. Thus, screening leprosy contacts through serology in areas of high endemicity can be useful in leprosy control measures (NAGAO-DIAS et al., 2007; SASAKI et al., 2001; SCOLLARD et al., 2006).

Motor function assessment should be carried out mainly in patients undergoing treatment, with the aim of early detection of disabilities. Well-defined sequelae can be found as early as diagnosis, such as unilateral or bilateral peripheral facial paralysis or paralysis of the orbicular branch of the zygomatic nerve (BRITTON; LOCKWOOD, 2004).

The identification of *M. leprae* is difficult due to the bacillus' *inability* to grow "*in vitro*", so the diagnosis of leprosy is based on microscopic detection of the bacillus in combination with clinical evaluation. However, bacilloscopy has low sensitivity as it requires at least 10,000 bacilli per gram of tissue for detection and comparing clinical classification, through the number of skin lesions, with the results of bacilloscopic examinations has shown that the clinical method also has low sensitivity (BUHRER-SEKULA et al., 2000; CROFT; CROFT, 1998).

The study of DNA using the PCR technique has been used for the molecular diagnosis of *M. leprae,* as it allows the presence of the bacillus to be detected even if it is in small quantities, suggesting an advantage when compared to conventional histopathological and baciloscopic examination (BANERJEE et al., 2010; CARDONA- CASTRO; BELTRAN-ALZATE; MANRIQUE-HERNANDEZ, 2008; JOB et al., 2008). PCR has a specificity of 100% and a high sensitivity (80%) when compared to sputum smear microscopy and histopathology (28% and 31% respectively). These differences in sensitivity between PCR and bacilloscopy, PCR and histopathology were statistically significant ($p < 0.001$ in both cases) (BANERJEE et al., 2010; KUMAR; DOGRA, 2009).

The development of PCR represented an opportunity for sensitive, specific and rapid detection of *M. leprae* (SCOLLARD et al., 2006). In the literature there are several target sequences for PCR and specific DNA probes for *M. leprae*, such as the genes encoding the 36-kDa antigen (DE WIT et al., 1991b; YOON et al., 1993), 18-kDa antigen (WILLIAMS et al., 1990; YOON et al., 1993), 65-kDa antigen (PLIKAYTIS; GELBER; SHINNICK, 1990; YOON et al., 1993) and *M. leprae* repetitive sequences (WOODS; COLE, 1990; YOON et al., 1993). Most have shown a high sensitivity of PCR, as from one to one hundred microorganisms per sample have been detected by this method (DONOGHUE; HOLTON; SPIGELMAN, 2001).

Several studies using skin biopsies from leprosy patients have shown that the PCR technique has greater sensitivity in detecting the bacillus, suggesting an advantage when compared to conventional histopathological and baciloscopic examinations (CARDONA-CASTRO; BELTRAN-ALZATE; MANRIQUE-HERNANDEZ, 2008; DONOGHUE; HOLTON; SPIGELMAN, 2001; JOB et al., 2008; KUMAR; DOGRA, 2009). Additionally, considering that PB types carry few microorganisms and that none of these cases would be detected by microscopic analysis, the positivity of PCR in PB forms (4.16%) demonstrated by Pontes et al. (2008) has become more significant, clearly showing an advantage over microscopy. Numerous clinical samples can be used to detect *M. leprae* using PCR, including cell scrapings, blood, skin and nerve biopsies and nasal secretions.

2 BACKGROUND

Despite the many advances made in the diagnosis of leprosy, the existence of regions with an agglomeration of cases and indications of active transmission (the North, Center-West and Northeast regions), which concentrate 53.5% of the cases detected in only 17.5% of the Brazilian population, living in extensive geographical areas, adds further complexity to effective interventions. The social and historical determinants associated with the occupation of the Legal Amazon and the maintenance of social inequalities in the Northeast region, help to explain the accumulation of infected people, in the case of a disease with a long incubation period (MINISTÉRIO DA SAÙDE-MS, 2013; KERR-PONTES et al., 2006). It is therefore necessary to intensify epidemiological surveillance in these areas, to monitor contacts, to maintain effective actions to stabilize the endemic, such as curing all cases diagnosed early and assessing and monitoring the physical disabilities presented by cases diagnosed late. But this will only be possible by expanding access to opportunities for diagnosis, treatment and surveillance.

Surveillance of contacts or individuals from endemic regions, who are possible carriers of *M. leprae* in low concentrations, is extremely important, as continued exposure can result in an increased risk of infection caused by active invasion by bacilli or by passive entry, resulting from a gap in the epithelium of the skin or nasal mucosa after physical trauma (PATROCÌNIO et al., 2005). In addition, more research is needed to confirm whether the DNA amplified in PCR from patients or contacts is really *M. leprae,* to understand whether the nasal reservoir is persistent, transient or recurrent and whether it induces protective immunity (CARDONA-CASTRO; BELTRAN-ALZATE; MANRIQUE-HERNANDEZ, 2008). Thus, the study of the detection of *M. leprae* DNA in the nasal secretions of patients and healthy individuals in an endemic area using the PCR technique, which allows *M. leprae to be* monitored at low concentrations. *leprae* at low concentrations, is relevant for the detection of asymptomatic carriers, as an auxiliary tool for the diagnosis of the disease, avoiding the use of other invasive diagnostic methods, for the elucidation of the site of exposure to the bacillus and the possible route of entry of the bacillus into humans. These studies would complement studies on the transmission of leprosy and intra- and extra-familial risk factors.

In addition, it is important to study the molecular epidemiology of leprosy, the genetic diversity of *M. leprae from* a local and global perspective, the distinction between cases of leprosy reactivation and reinfection, which can be done through genotyping systems. Genotyping methods are the only means by which leprosy transmission chains within populations can be established and patterns for genomic markers can be experimentally evaluated in the patient population of interest. This information provides a more detailed picture of the clonal structure of *M. leprae* strains circulating in patients in the state of Ceará, using nasal secretion samples, which are collected in a less invasive manner. In addition, the grouping derived from genotyping methods serves as a proxy for epidemiological kinship, even before contact tracing is carried out. However, it must be stressed that a more accurate description of the leprosy transmission scenario is provided by synthesizing both molecular and epidemiological data, contributing to the elimination of the disease as a public health problem.

3 OBJECTIVES

3.1 General Objective

To evaluate the genetic variability and spatial distribution among clinical isolates of *M. leprae* and to study the relationship between genetic composition and clinical-epidemiological factors.

3.2 Specific objectives

- To detect the presence of *M. leprae* DNA in the nasal mucosa of leprosy patients and correlate it with clinical and epidemiological factors.
- To detect the presence of *M. leprae* DNA in the nasal mucosa of healthy individuals and correlate it with clinical and epidemiological factors.
- To identify the most appropriate genetic markers for differentiating *M. leprae* strains in nasal secretion samples.
- To differentiate the strains of *M. leprae* present in nasal secretion samples among leprosy patients and correlate them with clinical and epidemiological data.
- To analyze the genetic polymorphism of *M. leprae* in the population of Fortaleza by analyzing isolates of nasal secretions from different patients.
- To analyze the difference between strains of *M. leprae* in nasal secretion and skin biopsy samples isolated from the same patient with leprosy.

4 METHODOLOGY

Two studies were carried out using different methodologies. The studies are being described separately for clarity. Both were carried out in the municipality of Fortaleza between November 2008 and December 2010 and were funded by the National Science Council (CNPq, project # 410573 / 2006-0).

The project "Clinical-epidemiological studies of Leprosy in an endemic area of the Brazilian Northeast", to which these studies are linked, was approved by the Research Ethics Committee of the Dona Libânia National Reference Center for Health Dermatology under protocol no. 011/07 (ANNEX A).

4.1 First Study

4.1.1 Type and place of study

This study was carried out at the Dona Libânia National Reference Center for Health Dermatology (CDERM) and the Christus School of Medicine, from November 2008 to December 2010. Initially, a case-control study was carried out, followed by a spatial distribution analysis study. Nasal secretions were collected from all study participants.

All participants who agreed to take part in the study signed an informed consent form, taking into account the provisions of Resolution No. 196 of October 10, 1996 (now Resolution No. 446) of the National Health Council. Patients under the age of 18 were asked to provide written parental consent.

4.1.2 Study participants

During the study period, all eligible cases and controls were approached and invited to take part in the study. These individuals were seen and examined by a highly trained dermatologist at the CDERM. This is a state administration unit, affiliated to the SUS, and a national center for Sanitary Dermatology.

A case of leprosy was defined based on the presence of one or more of the following criteria: a) typical skin lesion with loss of sensation; b) thickening of one of the main nerves with loss of sensation; c) positive bacilloscopy for *Mycobacterium leprae*. Diagnosis and classification were based on clinical characteristics and the bacilloscopic index (BI) (RIDLEY; JOPLING, 1966). The case group (C) consisted of 185 individuals diagnosed with leprosy, with no previous treatment, relapse cases and absentees who had returned to treatment provided that, in the latter two cases, the last dose was more than 5 years before the current restart.

The control group, known as Co, was made up of 136 individuals who sought dermatological services at CDERM for skin problems other than leprosy and who lived in Fortaleza. To avoid misclassification, these controls were examined to rule out leprosy. All eligible recruits were informed of the objectives and procedures of the study and signed an informed consent form.

For the spatial analysis, in addition to analyzing the spatial location of the cases and controls, a third group of participants was included, called the external group (EG). This group consisted of 121 individuals who, at the time of the study, lived in a geographic area with a high socioeconomic level and were different from the cases and controls. They were chosen from students at a private medical and physiotherapy school, in their second and third semesters, who did not attend the CDERM. The aim was to examine the prevalence of *M. leprae* in nasal secretions, as well as its geographical distribution, among cases, controls and a group (GE) that was very different in socio-economic terms from the other two groups, but all living in the same city considered to be endemic for leprosy (detection coefficient = 24.4 /100,000).

4.1.3 Data collection

Nasal secretions were collected from both nostrils of each participant. Samples from groups C and Co were collected at CDERM from November 2008 to December 2010. Samples and data from the EG were collected in June 2011 at the Christus College.

The participants in the case-control study (groups C and Co) were given a questionnaire (APPENDIX A) at the CDERM to record their home address and data on clinical, demographic and socioeconomic aspects.

In the GE group, the same questionnaire was applied at the Christus College, but reduced, containing only information on home address and demographic data such as gender and age.

4.1.4 Laboratory procedure

4.1.4.1 Collection and packaging

Nasal secretions were collected from both nostrils of each participant, using one swab for each nasal cavity. The swabs were previously moistened in TE buffer (Tris 10mM - EDTA 1mM, pH 8.0). The samples were collected by rotating them at the external end of the nasal orifice and then kept at -20°C until they were taken to the Mycobacteria Research Laboratory in the Biomedicine Block of the Federal University of Ceará (UFC) for DNA extraction.

4.1.4.2 DNA extraction

The tip of the swab was cut off with a sterile scalpel and transferred to a 2.0mL *eppendorf* tube along with the remaining liquid. 1mL of 4% NaOH was added and centrifuged at 2500 rpm for 20 minutes. All the supernatant was discarded and 600μL of S.E.B. solution (IOmM Tris-HCl, 10mM

EDTA, 50mM NaCl, 2% SDS and 0.3 mg/mL proteinase K) was added to the precipitate and incubated for 18 hours at 56°C. Then 600μL of phenol-chloroform-isoamyl alcohol (25:24:1) was added. The sample was then centrifuged at 12,000 rpm for 15 minutes, the supernatant was transferred to another tube and 500μL of chloroform was added, homogenized and centrifuged at 12,000 rpm for 5 minutes. Next, 35μL of 2.5mM NaCl and 500μL of isopropyl alcohol were added, and again centrifuged at 14,000 rpm for 20 minutes at 4°C. After this, the supernatant was discarded and 500 μL of 70% ethanol was added to the precipitate. The sample was then centrifuged at 14,000 rpm for 5 minutes, the supernatant discarded and the DNA resuspended in 70 μL of milliQ water (distilled and autoclaved).

4.1.4.3 Nucleic Acid Amplification

The *primers* were designed with the program *Primer3* v.0.4.0 (ROZEN; SKALETSKY, 2000) from the genomic region of *M. leprae* X17151 (*M. leprae repetitive elemental*, RLEP2).

Three primers (TABLE 01 and FIGURE 01) and two PCR reactions were used to amplify the 238 bp genomic fragment specific to *M. leprae* in 442 nasal secretion samples, totaling 884 PCR reactions.

The *Illustra™ Pure Taq Ready-to-go PCR Beads* kit (GE Healthcare) and the primers RLEP2.1 and RLEP2.2 (Invitrogen) were used for the first PCR and the primers RLEP2.2 and RLEP2.3 (Invitrogen) for the second PCR. For the first PCR reaction, 5 μL of extracted DNA, 1 μL of the RLEP2.1 primer, 1 μL of RLEP2.2 and 18 μL of milliQ water were used and for the second PCR reaction, 0.5 μL of the product of the first PCR, 1 μL of the RLEP2.2 primer, 1 μL of RLEP2.3 and 22.5 μL of milliQ water were used. The samples were taken to the *Gene Amp PCR System* 9700 thermal cycler (*Applied Biosystems*) and the two reactions were carried out as follows: initial denaturation at 94°C for 5 minutes, followed by 40 cycles of: 94°C for 30 seconds for denaturation, 59.5°C for 30 seconds for annealing and 72°C for 1 minute for extension. After the 40 cycles there was a final extension period at 72°C for 14 minutes and then the samples were kept at 4°C until they were removed from the thermal cycler and stored at -20°C. The amplification scheme for the second reaction was basically the same, changing only the annealing temperature to 59.8°C. The product generated by the first PCR was 282 bp, and the product generated by the second PCR was 238 bp.

Table 01 - Initiators for amplifying *M. leprae* DNA.

Initiators	Sense	Sequence (5' - 3')	Tm (°C)
RLEP2.1	Positive	ATATCGATGCAGGCGTGAG	59,80

RLEP2.2	Negative	GGATCATCGATGCACTGTTC	59,04
RLEP2.3	Positive	GGGTAGGGGCGTTTTAGTGT	60,24

Source: Author (2013)

Figure 01 - Amplification product from the *M. leprae repetitive element* RLEP2 genomic region (NCBI X17151).

```
atatcgatgcaggcgtgagtgtgaggatagttgttagcgccgcgggtaggggcgttttagtgtgcatgtcatggccttg
aggtgtcggcgtggtcaatgtggccgcacctgaacaggcacgtccccgtgcacggtataactattcgcacctgatgttat
cccttgcaccatttctgccgctggtatcggtgtcggcggcttgttgaccggccctcagccagcaagcaggcatgccgccg
ggtgcagcagtatcgtgttagtgaacagtgcatcgatgatcc
```

Source: Author (2013). In yellow: primers RLEP2.1 (5'-atatcgatgcaggcgtgag-3') and RLEP2.2 (5'-ggatcatcgatgcacgttc-3'), 282 bp product; in green: primer RLEP2.3 (5'- gggtaggggcgttttagtgt-3'), 238 bp product. Arrows indicate direction of amplification (Primer 3 Software).

4.1.4.4 Visualization of the PCR product

Ten microliters of the amplified product from the second PCR were applied to a volume of 50 mL of 2.0% agarose gel (*Invitrogen*) prepared in TBE 1x (Tris Base 100mM, Boric Acid 100mM and EDTA 6.4mM), containing ethidium bromide at a final concentration of 0.5 µg/mL. The electrophoretic run was carried out at 80 V for around 45 minutes (*Electrophoresis Power Supply* EPS 600 - *Pharmacia Biotech*), after which the bands were visualized and scanned under ultraviolet illumination in a transluminator (*ImageQuant* 300 - *GE Healthcare*).

4.1.5 Statistical analysis of groups C and Co

Fischer's exact test and *Pearson*'s chi-square test were used to evaluate the association in a contingency table with the PCR positive/negative group for *M. leprae* DNA between C and Co. For the factors that showed a significant association, the *Odds Ratio* and Confidence Interval were estimated with a 95% confidence coefficient. The difference between positive and negative PCR in relation to the mean of quantitative variables was assessed using the *Student's t-test* or the *Mann-Whitney test,* depending on whether the distribution of the variable was normal or not, respectively.

In order to verify in a multidimensional context the factors that increased the chance of obtaining a positive PCR, logistic regression analysis was used as a function of the independent factors that showed a significant association at the 5% significance level with the positive PCR group.

4.1.6 Spatial analysis of groups C, Co and GE

To enable spatial analysis and mapping of the individuals in groups C, Co and GE, the address of each of them at the time of entry into the study was geocoded by assigning a latitude and longitude to the address, using *Google Earth* software. The coordinates were transcribed into an Excel spreadsheet and transformed into PDF files to be used directly in the *GIS software ArcGIS 9.3*. After the data had been formatted into spreadsheets, the *ArcMap* application was used to make the information available on maps showing the distribution of the georeferenced individuals in Fortaleza according to the results of the experiments. Population densities were generated using the *Spatial Analyst - kernel* density extension, using a search radius of 20 km and a scale of 2.5 kilometers.

4.2 Second Study

A cross-sectional study was carried out with NHS samples from 48 individuals and skin biopsy (SB) samples from 38 individuals from group C of the first study, whose NHS samples were positive for *M. leprae* DNA (FIGURE 2).

Figura 2- Flowchart of the number of samples used in the second study.

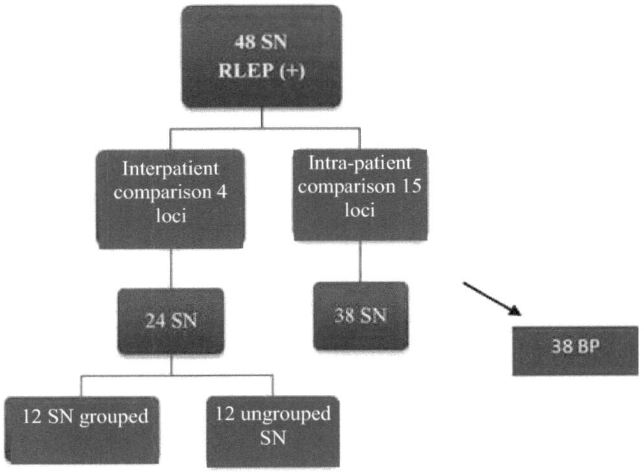

48 SN RLEP (+)
Interpatient comparison 4 loci
Intra-patient comparison 15 loci
24 SN
38 SN
38 BP
12 SN grouped
12 ungrouped SN

Source: Author (2014).

4.2.1 Skin biopsy samples

4.2.1.1 Collection and packaging

Samples of skin lesion biopsies were taken from 38 individuals in group C of the first study using a 5 mm *punch,* and this procedure was carried out by CDERM doctors. The *punch* blade was inserted perpendicular to the skin, with the axis of the instrument positioned at 90° to the plane of the skin surface. Rotating movements were made, starting the incision in a single direction, avoiding artifacts such as disintegration of fibers and attachments. The base of the tissue cylinder produced was incised by the *punch* blade itself, by tilting the handle to an angle of 3045°. The removed tissue was placed in a sterile 1.5 mL *eppendorf* which was stored ά - 20°C for subsequent DNA extraction.

4.2.1.2 DNA extraction

The skin biopsy samples were processed using the *DNeasy Blood & Tissue* genomic DNA extraction kit (*Qiagen*), following the manufacturer's instructions.

4.2.2 MLVA (*Multiple-locus VNTR Analysis*) of RLEP-positive nasal secretion and biopsy samples

To assess the genetic diversity of *M. leprae* between individuals, SN strains from 48 RLEP-positive patients in group C of the first study were analyzed using the four variable-number *tandem* repeat *loci* (VNTRs): AC8b, AC9, AC8a and GT9.

To study the variability of the strain in the same individual, BP and SN samples from 38 RLEP-positive patients in group C of the first study were compared using the fifteen VNTR *loci*: AT17, GGT5, GTA9, AC8b, AC8a, AT15, AC9, 21-3, GAA21, TA18, 6-7, 27-5, TA10, 23-3, 12-5.

The VNTRs, present in pseudogenes or intergenic segments of the bacillus genome, were separated in four multiplex PCR combinations (TABLE 2), using fluorescently-labeled 5' *forward primers*. The *primers* are specific to *M. leprae*, with the exception of (TA) 18 which can eventually produce a 242 bp cross-reaction product from human DNA (JENSEN et al., 2011; KIMURA et al., 2009).

Table 2- Initiators for amplifying 15 VNTR *loci* in multiplex PCR groups.

Locus	Forward Primers	Reverse primers	pb	PCR*
(AC)8b	VIC-GCCCACTTACCTCAACCAAC	CCTATAACGGCACTCAGTCCA	390	1
(GTA)9	NED-AGCCTTAGTCGCAGATG	TCCGCTGTCCGTCCGCTGA	307	
(GGT)5	6FAM-GCAGCGGTGTAACAGCATAGC	TGTCTGCCTTGCGAAACGGT	242	
(AT) 17	PET-TCTCCAACATGCTGCGACA	GTACAGCGGCCTGATCGAA	181	
21-3	6FAM-GAATCTGACCTTTCGGAAATG	CGATGCAGCTTCCTACGG	312	2
(AC)9	NED-AGCGCCCGTTGTCGATAGA	GACTGGATGTCGGCACCCC	236	
(AT) 15	PET-CAATATGCGGGTTGGCGCTTCTG	CCGTCTGGCTCGATGGCTGGATTC	168	
(AC)8[a]	VIC-GTGTTACGCGGAACCAGGCA	CCATCTGTTGGTACTACTGA	124	

27-5	6FAM-ATTGAGCAGATGGCCGGTC	AGCAGTCGGCACGCCCTT	327	
6-7	VIC-GCCATCGTTGTCGGTTCATC	CGGAGGAGGTGGGTACGGT	268	3
(TA) 18	NED-CGTGCGTCGTGTAGGC	GACGTGGCAACATCGAAGTT	230	
(GAA)21	PET-CTACAGGGGGCACTTAGCTC	GGACCTAAACCATCCCGTTTT	201	
12-5	VIC-CTGGTCCACTTGCGGTACGAC	GGAGAAGGAGGCCGAATACA	289	
23-3	6FAM-CCGAAGCCCTGGACGAAG	GCCGTAAATCCGCTCCC	243	4
(TA) 10	PET-TAGATTCAAACGACCATGCA	TGATAATCACGTGTTTCCGC	185	

Source: Jensen et al. (2011). *PCR: Multiplex PCR combination number.

4.2.2.1 Amplification conditions: multiplex PCR

Multiplex PCR was carried out using the *Qiagen* Multiplex PCR Kit (*Qiagen*). Each reaction contained 20 μL of final volume, made up of 10 μL of 2x *Qiagen* Multiplex PCR *Master Mix*, 2 μL of 5x Q solution, 2 μL (each) of *forward* and *reverse primers* and 2 μL of DNA; the volume was adjusted with sterile distilled water. The final concentration of each *primer* was 0.2 μM.

The following PCR conditions were used to obtain the amplified products: 15 min for initial denaturation at 95° C; then 40 cycles of denaturation for 30 s at 94° C, annealing for 90 s at 60° C, extension for 90 s at 72° C plus final extension at 72° C for 10 min.

4.2.2.2 Agarose gel electrophoresis

Ten microliters of the amplified product were applied to a volume of 50 mL of 2.0% agarose gel (*Invitrogen*) prepared in TBE 1x (Tris Base 100mM, Boric Acid 100mM and EDTA 6.4mM), containing ethidium bromide at a final concentration of 0.5 μg/mL. The electrophoretic run was carried out at 80 V for around 45 minutes (*Electrophoresis Power Supply* EPS 600 - *Pharmacia Biotech*), after which the bands were visualized and scanned under ultraviolet illumination in a transluminator (*ImageQuant* 300 - *GE Healthcare*).

4.2.2.3 Fragment analysis

After multiplex PCR, 1 μL of the PCR product was diluted 30-60 times, and 1 μL of the diluted PCR product was combined with 8.7 μL deionized formamide (*Applied Biosystems*) and 0.3 μL of LIZ-500 (*Applied Biosystems*). The material was denatured at 94° C for 5 min and subjected to capillary electrophoresis in the ABI 3130 DNA automatic sequencer (*Applied Biosystems*), injected into the capillary (length 50 cm, POP-7 polymer) by applying a voltage of 1 kV for 22s. Capillary electrophoresis was carried out at a voltage of 15kV at 60°C for 45min (JENSEN et al., 2011).

The electropherograms were visualized and analyzed, and the repeat numbers of the 14 VNTR *loci* were determined using *Peak Scanner*™ Software, version 1.0 (*Applied Biosystems*).

4.2.3 Calculation of Allelic Diversity

Allelic diversity (h) at each VNTR locus was calculated as recommended by Gilmour and Whittam (1986), using the formula: $h = 1 - \sum x_i^2$ [n / (n-1)]), where x_i is the frequency of allele i at the locus, and n is the number of isolates. Allelic diversity (h) is a good indirect indicator of sample heterogeneity/polymorphism and also a good indicator of the discriminatory power provided by the *loci* under study (GILMOUR; WHITTAM, 1986). According to Sola et al. (2003), based on this index, the sample is considered highly discriminatory/polymorphic if $h \geq 0.6$; moderately discriminatory if $0.3 \leq h \leq 0.6$ and weakly discriminatory if $h \leq 0.3$ (SOLA et al., 2003).

4.2.4 Calculation of discriminatory power

The discriminatory power of a typing method is its ability to distinguish between unrelated strains. This is determined by the number of types defined by the method and the relative frequency of the types (HUNTER; GASTON, 1988). The Hunter-Gaston discriminatory index (HGDI) can be used as a numerical parameter of the discriminatory power of RFLP, MIRU-VNTR Spoligotyping, alone or in combination. The HGDI was calculated as recommended by Hunter and Gaston (1988) using the formula in figure 3.

Figura 3- Equation for calculating the discriminatory power of molecular techniques

$$D = 1 - \frac{1}{N(N-1)} \sum_{j=1}^{s} x_j(x_j - 1)$$

Source: Hunter and Gaston (1988). D is the discriminatory power index, N is the number of unrelated strains tested, S is the number of different types, and xj is the number of strains belonging to type j, assuming that the groupings will be classified into exclusive categories (HUNTER and GASTON, 1988). Hunter and Gaston's equation is an application of Simpson's diversity index (SIMPSON, 1949).

4.2.5 Statistical analysis

The difference between patients (in group C) with strains with similar genotypes and unique genotypes in relation to the mean of quantitative variables was evaluated using the *Mann-Whitney* test.

4.2.6 Spatial analysis

The spatial analysis and mapping of leprosy cases (C) with strains with similar genotypes and unique genotypes were carried out as previously described for the first study.

5 RESULTS

Of the study participants, there was a predominance of females in the Co group, while there was no prevalence of males in the C group ($p < 0.0001$) (TABLE 1).

Table 1 - Study participants according to gender.

	Group C	Co	
Sex	N (%)	N (%)	OR (95% CI)
Male	99 (53.5)	33 (24.3)	3,593 (2,208 ; 5,847)
Female	86 (46.5)	103 (75.7)	1,000
Total	185 (100.0)	136 (100.0)	

Source: Author (2014). Fisher's exact test, $p < 0.0001$. OR, *Odds Ratio.*

Group C's age ranged from 4 to 81 years, with an average of 43, while group Co's ranged from 4 to 69 years, with an average of 26.5.

According to sputum smear microscopy and the *Ridley-Joplin* clinical classification, of the 185 cases in the study, 89 were classified as dimorphic, 37 as tuberculoid, 44 as virchowian, 9 as indeterminate and 6 did not have their clinical form informed.

The results of the two studies are being presented separately: results of the first study and the second study.

5.1 Results of the first study

In the amplification of the *RLEP* genomic region of *M. leprae* in nasal secretion, both groups C and Co of the CDERM had similar results (69.2%, 66.9%, respectively) (TABLE 2). The GE group had 34 (28.1%) positive and 87 (71.9%) negative individuals for *M. leprae* DNA.

Table 2 - Distribution of the results of the detection of the *RLEP* gene of *M. leprae* in groups C and Co

	Groups				Total		p
RLEP result	C		Co				
	n	%	n	%	n	%	
Positive	128	69,2%	91	66,9%	219	68,2%	0,716
Negative	57	30,8%	45	33,1%	102	31,8%	

Total	185	100,0%	136	100,0%	321	100,0%

Source: Author (2014). *Fischer*'s Exact Tests; C, cases; Co, controls.

For the logistic regression analysis of possible risk factors for a positive PCR result for the *RLEP* gene of *M. leprae*, groups C and Co were compared, since the EG had a constant profile of schooling and socioeconomic level, different from the other groups, and was only used for spatial analysis.

When the results of the *RLEP* that showed positivity to *M. leprae* in the two groups were analyzed in relation to gender, it was observed that the majority of the positives in the Co group were female, 71 out of 91 (78%), while in the C group, there was a predominance of males, 74 out of 128 (57.8%) (TABLE 3, $p < 0.0001$).

With regard to age group, it was observed that in those who tested positive for *RLEP,* the average age of group C was significantly higher than that of group Co (TABLE 3, $p < 0.0001$).

With regard to length of schooling (TABLE 3), in groups C and Co there was greater positivity for *RLEP* (58.6% and 40.7%, respectively) in individuals with less than 4 years of schooling (illiterate or with incomplete primary education) ($p = 0.0085$). Thus, in the regression analysis of possible risk factors for positive PCR for *M. leprae* in C and Co, it was observed that individuals with less than 4 years of schooling were more likely to have a positive *RLEP* compared to individuals with more schooling (TABLE 3, OR = 2.729).

With regard to the length of schooling of the head of the household (the main individual earning the family income), a higher level of positivity was observed in individuals in group C who had a head of household with less than 8 years of schooling, while in group Co a higher level of positivity was observed in individuals whose head of household had more than 8 years of schooling (65.8% and 55.7%, respectively, $p = 0.0028$). Thus, in the regression analysis of possible risk factors for *RLEP* positivity in C, it was observed that individuals whose head of household had less than 8 years of schooling were more likely to be PCR positive compared to those with more than 8 years of schooling.

While in the Co group, individuals whose head of household had more than 8 years of schooling were more likely to be *RLEP* positive than those with less than 8 years of schooling (TABLE 3, OR = 2.419).

In relation to the economic classification criteria of the RLEP-positive individuals (TABLE 3), greater positivity was observed in individuals in group C from levels D/E, while in group Co greater positivity was observed in individuals B/C (67.2% and 61.5%, respectively, $p < 0.0001$). Thus, in the regression analysis of possible risk factors for *M. leprae RLEP* positivity in C, it was observed that individuals in class D/E were more likely to be *RLEP* positive compared to those in class B/C. In the Co group, individuals from class B/C were more likely to be PCR positive than those from class D/E (OR = 3.276).

Among the RLEP-positive individuals, the frequency of the presence of a BCG scar was high in groups C and Co, 65% and 78%, respectively (p = 0.0460; TABLE 3). Thus, in the regression analysis of possible risk factors for a positive PCR result for *M. leprae* in groups C and Co, it was observed that individuals without the BCG scar were less likely to be PCR positive compared to those with the scar (OR = 0.512).

Table 3 - Logistic regression analysis for the association of socio-demographic factors between individuals in the C and Co *RLEP* positive groups

Variables	C N (%)	Co N (%)	Total	p	Odds ratio (95% CI)
Gender					
Male	74 (57.8)	20 (22.0)	94 (42.9)	<0.0001	4.865 (2.650, 8.932)
Female	54 (42.2)	71 (78.0)	125 (57.1)		1.000
Total	128 (100.0)	91 (100.0)	219 (100.0)		
Average age (sd)	40.9 (16.6)	30.6 (14.2)	36.7(16.4)	<0.0001	-
Education					
< 4 years	75 (58.6)	37 (40.7)	112 (51.1)	0.0085	2.729(1.436, 5.187)
4 to 8 years	27 (21.1)	19 (20.9)	46 (21.0)		1.913(0.881, 4.156)
> 8 years	26 (20.3)	35 (38.5)	61 (27.9)		1.000
Total	128 (100)	91 (100)	219 (100.0)		

Table 3 - Logistic regression analysis for the association of socio-demographic factors between individuals in the C and Co *RLEP* positive groups

Variables	C N (%)	Co N (%)	Total	p	Odds ratio (95% CI)
Education of the head of the					

family					
< 8 years	77 (65.8)	39 (44.3)	116 (56.6)	0.0028	2.419(1.37, 4.269)
> 8 years	40 (34.2)	49 (55.7)	89 (43.4)		1.000
Total	117 (100.0)	88 (100.0)	205 (100.0)		
Socio-economic level (Average gross income R$/month)					
B/C (6.006,00 - 3.118,00)	42 (32.8)	56 (61.5)	98 (44.7)	<0.0001	1.000
D/E (895,00)	86 (67.2)	35 (38.5)	121 (55.3)		3.276(1 870, 5.740)
Total	128 (100)	91(100.0)	219 (100.0)		
BCG scar					
Yes	80 (65)	69 (78.4)	149 (70.6)	0.0460	0.512(0.273, 0.961)
No	43 (35)	19 (21.6)	62 (29.4)		1.000
Total	123 (100)	88 (100.0)	211(100.0)		

Source: Author (2014). *Fischer*'s Exact Test; OR, *Odds Ratio*; CI, Confidence Interval with a 95% confidence coefficient. Economic classification, A to E according to the criteria of the Brazil Economic Classification Criteria points system of the Brazilian Research Association (http://www.abep.org/new/codigosCondutas.aspx).

When the PCR results in group C were analyzed according to sputum smear microscopy and *Ridley-Joplin* clinical classification, only sputum smear microscopy was statistically significant (TABLE 4, p = 0.0260), with *RLEP* positivity being higher in individuals who had a positive sputum smear microscopy (67 out of 85, 57.3%).

Table 4 - Distribution of PCR results for the *RLEP* gene of *M. leprae* in the group of cases according to smear microscopy and clinical classification according to *Ridley-Joplin*

PCR results								
	Positive		Negative		Total		p	OR
	N	%	N	%	N	%		(95% CI)
Bacilloscopy								

Positive	67	57,3	18	37,5	85	51,5	0,0260	2,233 (1,121; 4,451)
Negative	50	42,7	30	62,5	80	48,5		1,000
Total	117	100,0	48	100,0	165	100,0		
Clinical Classification								
Dimorphic	58	46,8	31	56,4	89	49,7	0,3746	-
Tuberculoid	27	21,8	10	18,2	37	20,7		
Lepromatous	34	27,4	10	18,2	44	24,6		
Undetermined	5	4,0	4	7,3	9	5,0		
Total	124	100,0	55	100,0	179	100,0		

Source: Author (2014). *Fischer*'s Exact Test ; OR, *Odds Ratio* ; CI, Confidence Interval with a 95% confidence coefficient.

Based on the results described in Table 5, in the multivariate context and at a significance level of 5%, it was observed that the factors age, gender and economic status influence the risk of presenting a positive *RLEP* result.

Thus, for an individual with a predefined socioeconomic level and gender, every year of age he gets implies that he has a 1.046 times greater chance of being *RLEP* positive. For individuals with a predefined age and socio-economic level, the fact that they are male increases their chance of being *RLEP* positive by 6.266 times compared to being female. It is also shown that if an individual is of a predefined age and sex, belonging to economic class D/E increases their chance of being *RLEP* positive by 3.083 compared to belonging to class B/C. Being young, of B/C socioeconomic status and female has a 0.081 times greater risk of being *RLEP positive* than the other factors (TABLE 5).

Table 5 - The factors that increase the chance of making group C in the logistic regression analysis in the PCR group positive in a multidimensional context

Factors	p	**OR**	**95% CI (OR)**	
			LI	**LS**
Age (years)	<0,0001	1,046	1,024	1,070
Male	<0,0001	6,266	3,079	12,748
Socio-economic level D/E	0,0010	3,083	1,600	5,940
Constant	<0,0001	0,081	-	-

Source: Author (2014). *Forward stepwise* method and *Wald* statistic. OR, *Odds* Ratio; CI, Confidence Interval with a 95% confidence coefficient.

The maps in Figure 4 show the location of the homes of the individuals in the three research groups.

Figura 4- Maps of the municipality of Fortaleza with the homes of the individuals in the study

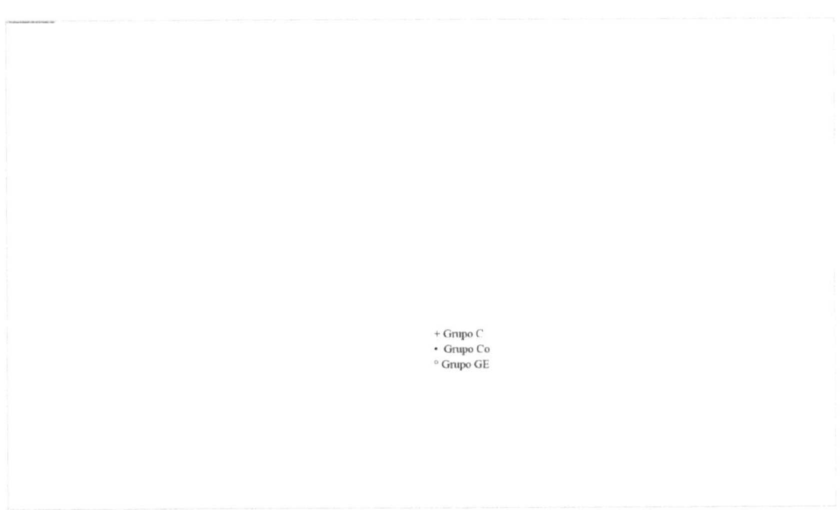

Source: Author (2014). A: *Kernel* density distribution of the homes of all the individuals in the three study groups (C, Co, GE). B: Location of the homes of all the individuals in the three study groups.

The maps in Figures 5 and 6 show the location of the homes of the individuals in the three groups in this study who tested positive for *M. leprae RLEP*, with Figure 6 showing the Kernel Density distribution. It was observed that groups C and Co are distributed throughout the municipality (FIGURE 5, maps B and C), however, groups C and Co show agglomerations (darker region of the map) on the outskirts of the municipality (FIGURE 6, maps B and C), regions that are less economically favored. The GE group, on the *other hand,* was observed to be clustered in a more economically favored region (FIGURE 6, map D). Map A in Figure 5 shows that group Co is the intersection between groups C and GE, as group Co meets group GE and group C, while group C has almost no geographical contact with group GE.

Figura 5- Maps of the Municipality of Fortaleza with the residence of the individuals by study groups

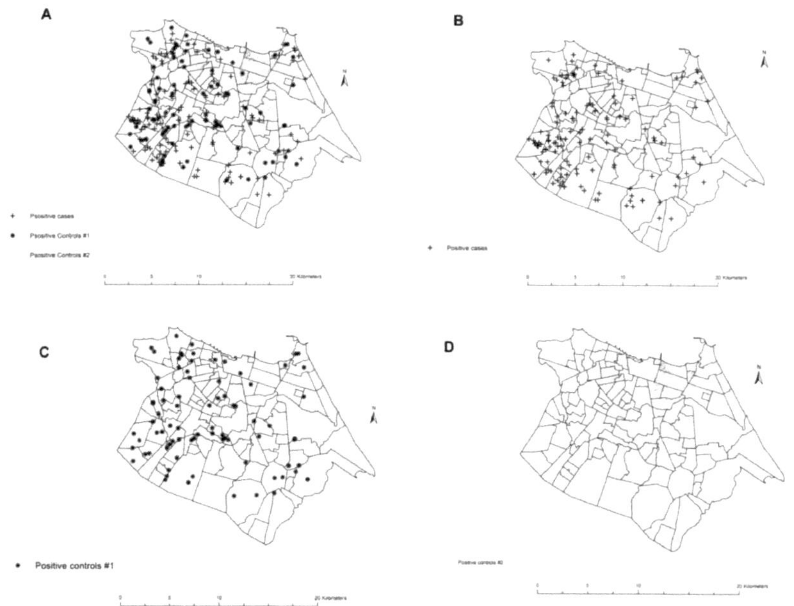

Source: Author (2014). A: home location of all the RLEP-positive individuals in the three study groups. B: home location of the RLEP-positive individuals in group C. C: location of the home of the RLEP-positive individuals in group Co. D: home location of the *RLEP-positive* individuals in the GE group.

Figura 6- Maps of the Municipality of Fortaleza in Kernel Density distribution of housing by study groups

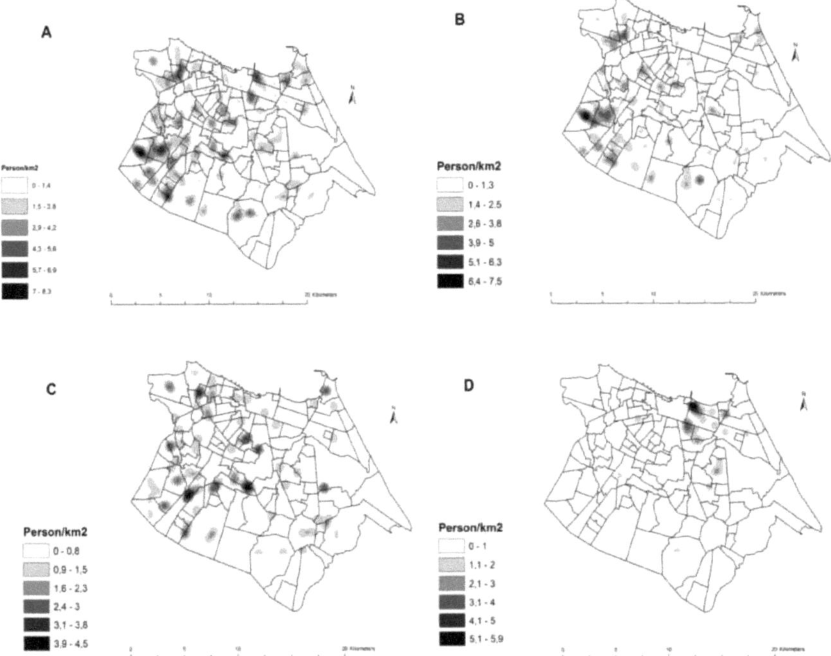

Source: Author (2014). A: home location of all RLEP-positive individuals in the three study groups. B: home location of the RLEP-positive individuals in group C. C: home location of the RLEP-positive individuals in group Co. D: home location of the RLEP-positive individuals in the GE group.

The map in Figure 7 shows the location of group C housing according to positive bacilloscopy and *RLEP* amplification results, in which it was observed that patients positive for bacilloscopy and *RLEP* are more clustered than patients positive for bacilloscopy and negative for *RLEP*.

Figura 7- Map of the Municipality of Fortaleza with the homes of group C individuals with positive bacilloscopy

Source: Author (2014).

5.2 Results of the second study

Nasal secretion (NS) samples positive for the RLEP gene from 48 individuals in group C were used to analyze the VNTRs published by, and skin biopsy (SB) samples from 38 of these individuals were also analyzed to study the variability of the strain in the patient.

Due to the behavior of some *loci* with little or no amplification, as well as high variability or invariability, only the AC8b, AC9, AC8a and GT9 *loci* were chosen for analysis of the SN samples and comparisons between patients. As shown in Table 6, variations in the number of copies were observed, ranging from six to eight for the AC8b locus with an allelic discrimination index (h) of 0.47, from seven to nine for the AC9 locus with h = 0.51, from eight to 11 for the AC8a locus with h = 0.57 and from eight to 16 for the GT9 locus with h = 0.57. Thus, the VNTR *loci* used in this study were considered moderately discriminatory.

Table 6- Allelic diversity of the 4 VNTR loci

VNTR	Number of alleles	Allelic diversity (h)	Conclusion	Total samples
67	8910 11	12 16		
AC 8b 1 30	17	0,47	Moderately discriminating	48
AC 919	191	0,51	Moderately discriminating	39
AC 8a	20 2031	0,57	Moderately discriminating	44
GTA 9	214 24	210,57	Moderately discriminating	43

Source: Author (2014).

With the loss of demographic data, of the total of 48 SN samples genotyped for the four *loci* (TABLE 7), only 24 samples were used for descriptive analysis and comparisons of the *M. leprae* genotypes defined in this study, of which four groups with identical *M. leprae* genotype strains were formed (TABLE 8), resulting in a Hunter - Gaston discriminatory index (HGDI) of 0.7273.

Table 7- VNTRs of the *M. leprae* strains from the 48 samples

Registration	AC 8b	AC 9	AC 8a	GTA 9
84267*	8	8	8	10
85.886	7	8	9	9
86187*	8	7	8	10
86.190	7	8	9	9
86.340	7	7	8	9
86.362	7	8	9	9
86.394	7	7	8	-
86.498	7	7	8	16
86.555	7	8	9	10
86610*	6	-	8	9
86.645	7	-	9	-
86776*	7	8	8	9
86.829	7	-	8	9
86.855	8	7	8	10
86962*	8	7	8	10

87126*	8	7	8	10
87184*	7	8	9	-
87212*	7	8	9	8
87.214	8	7	8	10
87330*	7	7	10	8
87401*	8	7	8	10
87.419	7	-	9	-
87.549	8	7	9	10
87.560	7	8	9	9
87914*	8	7	8	10

Table 7- VNTRs of the *M. leprae* strains from the 48 samples

Registration	AC 8b	AC 9	AC 8a	GTA 9
88.276	8	-	-	-
88689*	8	7	8	10
88742*	7	8	9	10
88935*	7	8	9	12
89.115	7	8	9	9
89160*	7	8	8	10
89167*	7	8	9	9
89.167	7	8	9	9
90148*	8	7	9	10
90373*	7	7	9	10
90.730	8	-	10	-
90.780	8	-	-	10
91.371	8	-	-	10
91.796	7	8	9	9
91845*	7	8	10	10
91.979	8	7	9	10
92008*	7	8	9	10
92070*	7	9	11	10
92114*	7	7	8	10
92.275	7	-	-	10
92662*	7	8	8	9
92.642*	7	7	8	10
-	8	7	8	12

Source: Author (2014).* 24 samples used for descriptive analysis.

Table 8 - Groups of samples with identical *M. leprae* genotype

Registration	Group	AC 8b	AC 9	AC 8a	GTA 9
86.776	1	7	8	8	9
92.662	1	7	8	8	9
92.642	2	7	7	8	10
92.114	2	7	7	8	10
88.742	3	7	8	9	10
92.008	3	7	8	9	10
87.401	4	8	7	8	10
86.187	4	8	7	8	10
87.126	4	8	7	8	10
87.914	4	8	7	8	10
86.962	4	8	7	8	10
88.689	4	8	7	8	10

Source: Author (2014).

The patient characteristics obtained through the questionnaire were evaluated among the grouped strain samples and the information considered most likely to be associated with the individuals in each group was selected (TABLE 9).

Group 1 consisted of two individuals with lepromatous leprosy, BCG scar, aged 51 and 62, who were born in a town in the interior of Ceará called Aracati, and who had lived in Aracati for 12 years. Both said they had known someone with leprosy, but that they were not related. The date of diagnosis was one year apart.

Group 2 was made up of two individuals with dimorphic leprosy, BCG scar, aged 59 and 60, diagnosed in 2010, 4 months apart. They were born in different towns in the interior of Ceará and moved to Fortaleza in 1995 and 1970, respectively.

Group 3 consisted of two individuals aged 34 and 32, diagnosed one year apart. The first was born and lived in Fortaleza, had dimorphic leprosy and a BCG scar. The second was born in Sao Paulo and moved to Fortaleza in 1997. He had lepromatous leprosy and no BCG scar.

Group 4 was made up of 6 individuals aged between 12 and 71 years old, all of whom were diagnosed in 2009. In this group, four individuals had dimorphic leprosy and the two who had lepromatous leprosy were diagnosed in February 2009.

Table 9- Main characteristics of the individuals belonging to the groups formed

Registration	Group	Date diagnosis	ofForm clinic	City of birth	Date of birth	Age	BCG scar
86776	1	11/03/2009	Virchowian	Aracati/CE	29/11/1946	62	yes
92662	1	19/08/2010	Virchowian	Aracati/CE	08/10/1958	51	yes
92642	2	18/08/2010	Dimorphic	Missao Velha/CE	18/08/1950	60	yes
92114	2	29/04/2010	Dimorphic	Morada Nova/CE	16/04/1951	59	yes
88742	3	09/09/2009	Dimorphic	Fortaleza/CE	16/06/1975	34	no
92008	3	21/06/2010	Virchowian	Sao Paulo/SP	12/12/1977	32	no
87401	4	25/04/2009	Virchowian	Uruburetama/CE	23/05/1964	45	yes
86187	4	01/02/2009	Dimorphic	Canide/CE	01/08/1937	71	no
87126	4	01/04/2009	Virchowian	Capistrano de Abreu/CE	29/10/1955	53	no
87914	4	01/07/2009	Dimorphic	Fortaleza/CE	19/03/1966	43	no
86962	4	01/03/2009	Dimorphic	Fortaleza/CE	22/05/1995	13	yes
88689	4	28/08/2009	Dimorphic	Sao Joao da Se/PI	01/07/1997	12	yes

Source: Author (2014).

Analysis of the geographic distribution of the patient population based on residential address revealed no clustering of patients with similar genotypes, while among patients with unique genotypes, greater geographic proximity was observed, with greater clustering in the western region of the municipality (FIGURE 8).

Figure 8- Maps of the municipality of Fortaleza showing the homes of individuals with the genotypes

of the *M. leprae* strain analyzed

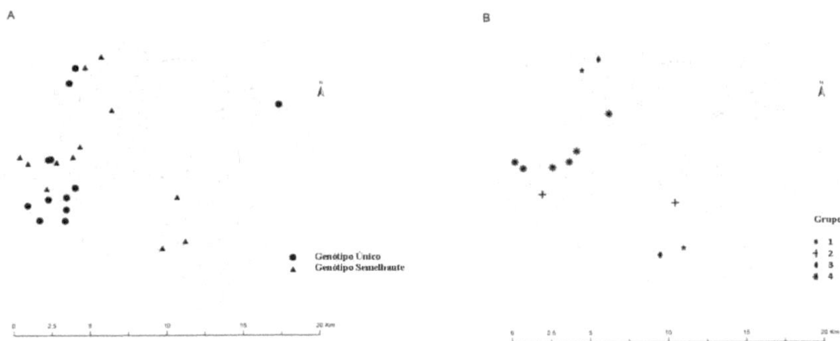

Source: Author (2014). A: Location of patients' homes according to those with single genotype strains and strains with similar genotypes. B: Location of patients' homes according to the groups formed from the strains with similar genotypes.

When some of the characteristics of the patients with strains with similar genotypes and those with unique genotypes were evaluated, in order to identify which ones would be associated with the formation of the groups, it was observed that the patients who were classified into groups had a mean age of 45 years, while those with unique genotypes had a mean age of 25 *(p* = 0.0110). The average time between symptoms and diagnosis for patients with grouped strains (similar genotypes) was lower than for those with a single genotype, but this was not statistically significant, while the baciloscopic index was the same for both groups of patients (TABLE 10).

Table 10- Comparison of the characteristics of patients with strains with similar genotypes and unique genotypes by the *Mann-Whitney* test

	Samples	N	Average	dp	Median	p
Age	Single genotypes	12	25,500	8,163	23,5	0,0110
	Similar genotypes	12	44,583	18,807	48,0	
Time (months) between symptoms and diagnosis	Single genotypes	12	22,625	17,391	24,8	0,7728
	Similar genotypes	12	20,304	17,505	13,1	
Bacilloscope index	Single genotypes	9	4,239	0,302	4,0	0,9670
	Similar genotypes	10	4,100	1,312	4,3	

Source: Author (2014). sd, standard deviation.

SN and BP samples from the same patient were analyzed to investigate the possibility of *M. leprae* genetic diversity in the individual. Samples from 38 patients were compared and 15 VNTR *loci were* evaluated (TABLE 11).

Some *loci of M. leprae* samples from SN did not amplify. Therefore, in the genotypic comparison analysis of the strains originating from the same patient, the quantities of divergent *loci* and *loci* that did not amplify were observed (TABLE 11).

Among the results, samples from five patients did not differ in any amplified *loci,* but the NS from two of these five patients did not amplify a *locus* and the NS strains from the other three patients did not amplify two *loci* (TABLE 11).

Another two patients had strains with an actual similarity of 86.7%, since all the *loci* amplified (TABLE 11).

Another pair of patients had strains that differed in seven *loci.* These patients had SN and BP with the most different strains, since in one patient all 15 *loci* were amplified and in the other only one *loci* did not amplify (TABLE 11).

Strains with notable differences between SN and BP were also observed in four patients. Their samples diverged at five *loci,* also with some *loci* not amplified. Another patient had strains that diverged at six *loci,* with only one *loci* not amplified. Additionally, it was observed that most of the *M. leprae* strains from the same patient with different VNTR *loci* involved increased copy number in the BP samples (TABLE 11).

Table 11- VNTRs of *M. Ieprae* strains from nasal secretion and skin biopsy samples from the same patient

Registration	Sample	AT17	GGT5	GTA9	AC8b	AC8a	AT15	AC9	21-3	GAA21	TA18	6-7	27-5	TAIO	23-3	12-5	№ VNTRs not amplified	№ Different VNTRs
86.555	SN	11	4	10	7	9	16	8	2	12	21	6	5	-	2	4	1	0
	BP	11	4	10	7	9	16	8	2	12	21	6	5	8	2	4		
87.212	SN	12	4	8	7	9	14	8	2	13	23	6	5	-	2	4	1	0
	BP	12	4	8	7	9	14	8	2	13	23	6	5	8	2	4		
87.156	SN	10	4	9	-	9	16	7	2	14	19	6	5	-	2	4	2	0
	BP	10	4	9	7	9	16	7	2	14	19	6	5	8	2	4		
86.190	SN	11	4	9	7	9	-	8	2	11	21	6	5	-	2	4	2	0
	BP	11	4	9	7	9	28	8	2	11	21	6	5	8	2	4		
87.401	SN	19	4	10	8	8	-	7	2	14	16	6	5	-	2	4	2	0
	BP	19	4	10	8	8	24	7	2	14	16	6	5	8	2	4		
87.914	SN	16	4	10	8	8	**16**	7	2	**14**	18	6	5	10	2	4	0	2
	BP	16	4	10	8	8	**15**	7	2	**15**	18	6	5	10	2	4		
91.796	SN	13	4	9	7	9	16	8	2	11	**25**	**7**	5	8	2	4	0	2
	BP	13	4	9	7	9	16	8	2	11	**21**	**6**	5	8	2	4		
86.187	SN	-	4	10	8	8	**22**	7	2	15	15	6	5	-	2	4	2	2
	BP	21	4	10	8	8	**23**	7	2	15	15	6	5	8	2	4		
85.886	SN	13	4	9	7	9	**17**	8	2	12	**15**	7	5	-	2	4	1	2
	BP	13	4	9	7	9	**18**	8	2	12	**17**	7	5	8	2	4		
88.935	SN	13	4	12	7	9	**18**	8	2	11	**18**	**7**	4	10	2	4	0	3
	BP	13	4	12	7	9	**19**	8	2	11	**19**	**6**	4	10	2	4		
89.167	SN	12	4	9	7	**9**	**15**	8	2	12	17	**5**	5	9	2	4	0	3
	BP	12	4	9	7	**10**	**14**	8	2	12	17	**6**	5	9	2	4		
92.070	SN	14	4	10	7	11	16	9	2	9	-	**7**	5	**10**	2	4	1	2
	BP	14	4	10	7	11	16	9	2	9	26	**6**	5	**8**	2	4		
87.126	SN	**22**	4	10	8	8	-	7	2	**13**	17	7	5	-	2	4	2	2
	BP	**12**	4	10	8	8	29	7	2	**14**	17	7	5	8	2	4		
89.160	SN	11	4	10	7	8	**16**	8	2	**11**	-	**7**	5	8	2	4	1	3
	BP	11	4	10	7	8	**15**	8	2	**12**	21	**6**	5	8	2	4		

Table 11- VNTRs of *M. Ieprae* strains from nasal secretion and skin biopsy samples from the same patient

Registration	Sample	AT17	GGT5	GTA9	AC8b	AC8a	AT15	AC9	21-3	GAA21	TA18	6-7	27-5	TAIO	23-3	12-5	№ VNTRs not amplified	№ Different VNTRs

90.148	SN	13	4	10	8	9	10	7	2	13	15	7	5	8	2	4	0	4
	BP	20	4	10	8	8	29	7	2	14	15	7	5	8	2	4		
86.855	SN	18	4	10	8	8	-	7	2	14	19	7	5	-	2	4	2	3
	BP	14	4	10	8	9	25	7	2	14	20	7	5	8	2	4		
84.267	SN	13	4	10	8	8	16	8	2	12	18	6	5	-	2	4	1	4
	BP	12	4	9	7	9	16	8	2	12	18	6	5	8	2	4		
87.851	SN	15	4	-	-	8	16	7	2	12	15	6	4	8	2	4	2	3
	BP	16	4	15	7	8	16	7	2	12	14	6	5	8	2	4		
88.742	SN	13	4	10	7	9	16	8	2	11	26	6	5	8	2	4	0	5
	BP	13	4	9	7	9	15	8	2	13	25	5	5	8	2	4		
88.689	SN	16	4	10	8	8	-	7	2	15	-	7	5	8	2	4	2	3
	BP	17	4	10	8	8	22	7	2	16	25	6	5	8	2	4		
86.776	SN	13	4	9	7	8	21	8	2	12	16	X	5	-	2	4	2	4
	BP	13	4	9	8	9	22	8	2	12	17	6	5	8	2	4		
87.214	SN	18	4	10	8	8	-	7	2	14	22	-	-	-	2	4	4	2
	BP	18	4	10	8	8	33	7	2	13	25	6	5	8	2	4		
87.250	SN	17	4	10	-	8	-	7	2	14	17	7	-	-	2	4	4	2
	BP	17	4	10	8	8	26	7	2	15	16	7	5	8	2	4		
87.370	SN	18	4	9	-	9	13	8	-	10	19	7	5	-	2	5	3	3
	BP	19	4	9	7	9	12	8	2	10	19	7	4	10	2	5		
86.962	SN	17	6	10	8	8	-	7	2	17	19	6	5	-	2	4	2	4
	BP	25	4	10	8	8	34	7	2	16	14	6	5	8	2	4		
87.669	SN	11	4	12	-	9	19	7	2	12	-	6	4	8	2	4	2	4
	BP	15	4	12	7	9	17	7	2	15	17	6	2	8	2	4		
88.053	SN	14	4	10	-	9	19	8	2	11	16	5	5	5	5	4	1	6
	BP	12	4	9	7	9	16	8	2	10	16	6	5	10	2	4		
86.340	SN	14	4	9	7	8	22	7	2	10	-	6	5	-	2	4	2	5
	BP	13	4	9	7	9	14	8	2	12	30	6	5	9	2	4		

Table 11- VNTRs of *M. Ieprae* strains from nasal secretion and skin biopsy samples from the same patient

Registration	Sample	AT17	GGT5	GTA9	AC8b	AC8a	AT15	AC9	21-3	GAA21	TA18	6-7	27-5	TAlO	23-3	12-5	№ VNTRs not amplified	№ Different VNTRs
91.845	SN	11	4	10	7	10	16	8	2	9	20	7	5	8	2	4	0	7
	BP	17	4	9	8	9	16	8	2	15	23	6	5	8	2	4		
91.979	SN	15	4	10	8	9	-	7	2	14	16	7	5	8	2	4	1	
	BP	13	4	9	7	9	16	7	2	11	14	6	5	8	2	4		
89.230	SN	-	4	7	-	9	-	8	2	11	20	-	-	8	2	-	6	2
	BP	12	4	9	7	9	16	8	2	12	20	6	5	8	2	4		
87.188	SN	18	4	10	-	9	-	7	2	14	-	-	-	-	2	3	6	3
	BP	20	4	10	8	8	28	7	2	14	17	6	5	8	2	4		
87.549	SN	-	4	10	8	9	-	7	2	-	22	-	4	-	-	-	7	3
	BP	11	4	10	8	9	16	7	2	13	14	6	5	8	2	4		
87.728	SN	12	4	10	-	9	18	-	2	10	-	5	-	11	2	-	5	5
	BP	18	4	10	8	8	25	7	2	19	23	6	5	8	2	4		
88.079	SN	16	-	-	-	8	-	-	-	14	13	6	4	8	2	4	6	3
	BP	16	4	10	8	8	30	6	2	13	5	6	5	8	2	4		
87.197	SN	18	4	9	-	9	-	7	-	14	-	7	-	-	-	3	7	4
	BP	18	4	11	8	9	21	7	2	15	10	6	5	8	2	4		
88.119	SN	12	4	13	-	10	15	7	2	10	-	-	6	-	-	-	6	5
	BP	16	4	7	7	10	20	7	2	14	18	7	5	8	2	4		
88.899	SN	11	4	11	-	-	20	-	-	14	-	-	-	9	2	-	8	3
	BP	13	4	9	8	8	22	7	2	14	19	6	5	9	2	4		

Source: Author (2014). In yellow the different loci, in red the largest numbers of different loci.SN=Nasal secretion; BP=Biopsy.

6 Discussion

6.1 First study

The detection of *M. leprae* by PCR has shown high specificity and sensitivity, being able to detect in nasal secretion samples a positivity of 100% for the MB clinical forms, 34 - 80% for the PB forms (SClOLLARD et al., 2006), 5 - 12% in MB contacts and 1- 2% in PB contacts (BANERJEE et al., 2010). Surveillance is extremely important in contacts of patients and in individuals from endemic regions, who are possible carriers of *M. leprae* (PATROCÌNIO et al., 2005). However, it is difficult to carry out due to the lack of specific laboratory tests for *M. leprae,* since this bacillus does not grow in axenic culture media.

In this study, the PCR technique was used to evaluate the frequency of this bacterium's DNA in nasal secretions of individuals, comparing leprosy patients and non-leprosy patients in Fortaleza, Ceará. Nasal secretions are used as a good alternative for surveillance of leprosy contacts, because not only are they not collected invasively, but the upper airways are the entry and exit points for the bacteria (SCOLLARD et al., 2006). Continued exposure of the individual to the bacillus can result in an increased risk of infection caused by active invasion of the bacillus or by passive entry, resulting from a solution of continuity in the epithelium of the nasal mucosa after some physical trauma.

In addition to the detection of carriers of *M. leprae* DNA, this study addressed for the first time that age, baciloscopic index and schooling can influence the positivity of the PCR for the RLEP of *M. leprae,* since the literature only describes factors that influence leprosy (ALMEIDA et al., 2004; BANERJEE et al., 2010; CUSTODIO et al., 2012).

In the current study, the detection rate of *M. leprae* DNA in group C was 69.2%, which is considered high compared to some studies in endemic areas, which reported a positivity range between 15% and 65% (PONTES *et al.*, 2008; BAKKER et al., 2006; CUSTODIO et al., 2012; DE WIT et al., 1991; KLATSER et al., 1993; PATROCÌNIO et al., 2005). Some of these studies used the RLEP region, but others used different regions such as the *pra* genomic region and simple PCR as the amplification method.

Several studies with contacts show a positivity rate ranging from 1.7% to 23% (ALMEIDA et al., 2004; BANERJEE et al., 2010; CUSTODIO et al., 2012; NAGAO- DIAS et al., 2007), much lower than in the current study, in which the Co group showed similar positivity to the C group (66.9%). This can be partly explained by the fact that our study used the *nested* PCR technique, which has greater sensitivity than the simple PCR used in other studies (ALMEIDA et al., 2004; BANERJEE et al., 2010; CUSTODIO et al., 2012; NAGAO-DIAS et al., 2007). Another explanation for the high positivity in the Co group is that these participants are treated in the same health service (CDERM) as the leprosy patients, group C. The CDERM is a place where low-income patients living on the outskirts of Fortaleza are treated, which are the most endemic areas for leprosy in the city (DATASUS, 2013). From the georeferencing, it was observed that individuals in groups C and Co live on the outskirts of the district or in less economically favored regions, predominantly in administrative regions I and V of the metropolitan region of Fortaleza.

These regions are already recognized by the Instituto de Pesquisa e Estratégia Econòmica do Cearà (IPECE) as having the lowest average personal income. However, it should also be noted that individuals in the Co group are also found scattered in other regions of the municipality, unlike the spatial distribution of the C group, which is concentrated in regions V, III and I. As such, contact between these two groups may have occurred in the neighborhoods where they lived or during occasional meetings at the CDERM, implying that the bacillus spread throughout the population.

The GE group had the lowest PCR positivity (28.1%) when compared to the other groups. This GE group was made up of young medical and physiotherapy students from a private college in the state of Ceará, thus representing a portion of the population living in administrative region II, a region containing the wealthiest neighborhoods in Fortaleza, including Meireles, Aldeota and Dionisio Torres. According to the IPECE, the Meireles neighborhood has an average personal income 15.3 times higher than that of a poor neighborhood in region V.

Although the GE group had a low positivity (28.1%) when compared to groups C and Co, it was significant when compared to the studies by Banerjee et al. (2010) who found positivity in 10.9% of MB contacts and 1.3% of PB contacts, Almeida et al. (2004) with positivity of 1.7% among intra-household contacts of patients, and Custodio et al. (2012) with positivity of 23.07% among contacts. Several studies are unanimous in stating that contacts of patients with positive nasal secretion PCR have a very high risk of developing leprosy (BAKKER et al., 2006; DOUGLAS et al., 2004). It should also be pointed out that the bacillus invades the nasal mucosa and reaches the blood vessels (PATROCÌNIO et al., 2005), after which it reaches the blood cells.

Schwann cells as well as the mononuclear phagocytic system. Most contacts will not develop the disease due to the complex relationship between genetic, immunological and environmental factors, but they are carriers and potential transmitters of the bacillus to susceptible individuals (GOULART; GOULART, 2009).

The detection rates found in groups C and Co and GE can be explained by the fact that a large part of the population in the poorer neighborhoods depends on the jobs generated in the richer neighborhoods, so there is contact between the different populations and the spread of the bacillus. According to Hausfeld (1970), the risk of being infected with the leprosy bacillus is not restricted only to household contacts and neighbors, but also to the entire structure of the community (HAUSFELD, 1970).

By analyzing the geographic distribution of the groups' homes, according to the percentage of positive nasal secretions, it was observed that group Co is distributed throughout the municipality and is the intersection between groups C and GE. The Co group meets the SG and the C group, while the C group has little contact with the SG. Thus, the Co group may be transmitting the bacilli from group C to GE. Since Fortaleza is a highly endemic area for leprosy (DATASUS, 2013), the population generally comes into contact with people who carry the bacillus, whether they are subclinical leprosy patients or healthy carriers, and this contact causes the microorganism to spread silently through the population. Similarly, in a study of nasal carriers in a village in Ethiopia, it was reported that healthy carriers and subclinical cases act in the bacillus transmission chain (BEYENE et al., 2003). Also in this Ethiopian study, in a geographical distribution analysis, it was observed that cases have a 90% probability of forming clusters, while the group of controls carrying *M. leprae form* clusters in the areas adjacent to the cases.

Also in the analysis of the geographical distribution of housing in group C, it was observed that patients with positive bacilloscopy and positive RLEP are more geographically clustered than patients with positive bacilloscopy and negative RLEP, which confirms the greater participation in the leprosy transmission chain of patients in whom *M. leprae* DNA was detected in the nose.

In relation to gender, group C had more men positive for RLEP than group Co. In the context of multivariate analysis, the fact that the individual is male has a 6.266-fold increase in the chance of being PCR positive, which reflects a greater probability of this gender acquiring the disease. In a cohort study carried out by Bakker et al (2006), it was shown that males have twice the risk of acquiring leprosy compared to women. One explanation may be due to the fact that men are more exposed to the environment than women, since men are more likely to work in professions where they come into contact with soil, water and animals. However, Klatser et al. (1993) suggest that men and women are equally exposed to the bacillus, with a higher proportion of men among the cases due to men being more susceptible to developing the disease after being infected. According to the World Health Organization, the proportion of men among new leprosy cases varies in different parts of the world, with 53.25% in the Americas and 68.8% in the Dominican Republic (WHO, 2013).

The mean ages in groups C and Co were different. In group C, this average was higher than in Co, which may be due to the long incubation period of the disease, which is usually between 2 and 7 years, leading to the ages of affected patients generally being over 40, affecting people of more advanced ages (KERR-PONTES et al., 2004). In a previous study carried out by our research group investigating anti-PGLI seropositivity in cases and contacts of cases, it was also observed that the average age of seropositive cases was higher (47 years) than that found in seropositive contacts and non-contacts of cases (FROTA et al., 2010a). However, this was also observed when dividing these groups in relation to the PCR result. Thus, in the RLEP PCR-positive individuals, the average age of group C is even higher than in group Co and in the context of multivariate analysis, each year that an individual ages implies that they are 1.046 times more likely to have a positive PCR for RLEP. This demonstrates early contact with the bacillus in this population, such as the Co group, who may be coming into contact with the bacillus early on and becoming ill in the future, becoming the C group, as demonstrated by Banerjee et al (2010). The controls' early contact with the bacterium may have occurred through a leprosy patient or through environmental contact. Studies indicate the presence of *M. leprae* in rivers, lakes and wells (FROTA et al., 2010b; MATSUOKA M et al., 1999), in the soil (LAVANIA et al., 2008) and in armadillo species (FROTA et aL., 2012; TRUMAN, 2005), which are native animals to the study region.

The highest PCR positivity rate was in patients with a positive bacilloscopy, since individuals with a high bacillary load (high BI) have a greater chance of testing for *M. leprae* DNA in nasal secretions. However, when analyzing group C, distributed according to the Ridley-Joplin clinical classification, there was no significant difference between positive and negative PCR for *M leprae,* with a high positivity in the tuberculoid (27 out of 37) and indeterminate (5 out of 9) forms, leading to the questioning of the role of these paucibacillary patients in the transmission of the bacillus in the region under study. It was expected to find greater positivity in the multibacillary forms, as in the study by Banerjee *et al* (2010) which found *M. leprae* DNA in 49.4% of nasal secretion samples from TT patients, 51.6% in DT, 62.5% in DV and 78.1% in V. A study carried out in Brazil (ARAUJO *et al.*, 2012) found positivity rates of 5.3% and 46.5% in PB and MB patients respectively.

The BCG vaccine, originally designed to protect against tuberculosis, has also been shown to be effective against leprosy in several studies (CONVIT et al., 1992; CUNHA et al., 2004; FINE et al., 1986; RODRIGUES et al., 2007; ZODPEY; AMBADEKAR; THAKUR, 2005). Most vaccinated seropositive contacts who develop leprosy show the milder form of the disease (PB) with negative serology for M. leprae (DÜPPRE et al., 2008; LOBATO et al., 2011). In the current study, no protective effect of BCG was observed against the colonization of *M. leprae* in the nasal mucosa, since there were more PCR-positive individuals who had the scar. A study by Beyene et al. (2003) also found no association between PCR positivity in nasal secretions and a history of BCG vaccination or scarring. This leads us to hypothesize that BCG does not play a protective role in the nasal mucosa. Since the bacillus most often enters through the upper respiratory tract, several studies have shown that a vaccine administered via the nasal mucosa produces more effective protection against tuberculosis, as it produces not only a systemic immune response, but also immunity in the mucosa of both the upper and lower airways (BROWN et al., 2003; CHEN et al., 2004; GLATMAN-FREEDMAN, 2006; GOONETILLEKE et al., 2003; HIROI et al., 1998; KÀLLENIUS et al., 2007). This could also be related to *M. leprae*.

Several studies have shown an agreement between areas with a higher incidence of leprosy and lower socioeconomic levels (CURY et al., 2012; DA CUNHA et al., 2007; KERR-PONTES et al., 2006; RODRIGUES et al., 2007; SALES et al., 2011). In agreement with the literature, this study also found an increased chance of 3.083 of the presence of *M. leprae* DNA in the nasal mucosa of individuals belonging to socioeconomic class D/E.

Lower socio-economic levels are related to lower education levels. Thus, in this study, the illiteracy of the individual and the education of the head of the individual's family of less than eight years were associated with an increased chance of positivity for RLEP in nasal secretion (OD = 2.729 and OD = 2.419, respectively). Low schooling has *already* been reported as a risk factor for leprosy in a case-control study carried out by Kerr-Pontes *et al (*2004), in which they obtained an OR of 1.81.

PCR is currently widely used in research, as it is a fast and accurate method for detecting *M. leprae* DNA in various clinical samples. It can help to classify patients, complementing the clinical and histopathological diagnosis of the disease, in order to lead them to appropriate treatment regimens, reducing the risk of disability and transmission of the disease. Thus, using PCR, this study showed a high positivity rate in the three groups studied, demonstrating several individuals colonized with *M. leprae* in the upper airways who may be playing an active role in the transmission of leprosy, and risk factors related to the detection of *M. leprae* DNA in nasal secretions.

6.2 According to the study

One difficulty in studying the molecular epidemiology of leprosy is the homogeneity of *M. leprae* DNA. The *M. leprae* genome is very conserved and genetic variations are exceptionally rare. However, these variations are attributed to the loss of genes that normally function to ensure replication fidelity, such as *mut* or *dnaQ* genes, as well as VNTR regions (PHETSUKSIRI et al., 2012).

Thus, several researchers have investigated polymorphic markers within the *M. leprae* genome and have evaluated their potential as epidemiological tools for differentiating strains (GILLIS et al., 2009; PHETSUKSIRI et al., 2012; SAKAMURI et al., 2009; ZHANG; BUDIAWAN; MATSUOKA, 2005).

This study used the method of Kimura et al. (2009) and Jensen et al. (2011) to assess the genetic variability of *M. leprae* isolates in Fortaleza, Ceará. This study was one of the first to use SN for genotyping, and compared a greater number of *loci* and samples than those used in previous studies (MATSUOKA et al., 2004a; YOUNG et al., 2004).

The genotyping of *M. leprae* is more complex than that of other bacteria due to its long incubation period, low pathogenicity and the fact that DNA from this bacillus can only be obtained from clinical specimens from patients due to the impossibility of *in vitro* cultivation. Thus, the SN, the sample chosen in this study, was easy to collect, represents the entry and exit point for the bacterium and has rarely been used for genotyping in previous studies, but it has low viability and contamination with other microorganisms. Also, the *M. leprae* DNA extracted from this material is of low quality and in quantities that are often insufficient for carrying out the experiments.

The difficulty in analyzing VNTRs in NS samples in the current study was also observed in the work by Young et al (2004). Although these authors studied only three *loci in* samples from five different sites in the body of patients who were collected to assess the genetic diversity of *M. leprae in* a single individual, only in the samples of the SN and peripheral blood mononuclear cells were all the *loci* not amplified, and even less amplification was shown in the SN samples. In addition, other studies have also discussed the lack of amplification in purer samples compared to NS. As in the study by Da Silva Rocha et al (2011) who used 4 VNTRs (GAA, GTA9, AT17 and TA18) to differentiate unrelated strains of *M. leprae and*, even using biopsy samples, were unable to amplify all four *loci* in all samples (DA SILVA ROCHA et al., 2011).

In addition to the complications related to the methodology, in this study some patient information was incomplete, which made it difficult to analyze the results obtained.

The evaluation of a typing method is based on three factors: typing capacity, reproducibility and discrimination. Discriminatory power is the ability of the method to distinguish between unrelated samples and this is determined by the number of types defined by a method and the relative frequencies of these types (HUNTER and GASTON, 1988).

Hunter and Gaston (1988) suggested the use of a numerical discrimination index, based on the probability of two uncorrelated samples from a test population being identified as belonging to different groups. This probability is calculated by the Hunter and Gaston Discrimination Index (HUNTER; GASTON, 1988), which was developed from Simpson's Diversity Index (SIMPSON, 1949) for describing species diversity in a given ecological habitat. According to the authors, a test for epidemiological studies is considered efficient when it has a discrimination index of at least 90%. In this study, an HGDI of 73% was obtained with four VNTR loci, but only the 24 samples with complete demographic data were used for this calculation.

The occurrence of wide genetic variations between strains based on the use of hypervariable repeats points to the need for caution when using highly dynamic markers. Thus, the four loci used for analysis of the NS samples and comparisons between patients have already been well studied in terms of their stability and polymorphic potential (GROATHOUSE et al., 2004; KIMURA et al., 2009; TRUMAN et al., 2004; YOUNG et al., 2004) and in this study showed an allelic diversity considered to be moderately discriminating.

In the current study, AC8b had an allelic diversity (*h*) of 0.47, unlike another study carried out in Brazil with samples from Sao Paulo and Rio de Janeiro, in which this locus had $h = 0.05$ (FONTES et al., 2009). In our study, this locus showed six to eight repetitions, with seven being the most frequent, which was also seen in another study carried out in Brazil (FONTES et al., 2009). In Thailand (SRISUNGNAM et al., 2009), seven alleles were also the most frequent, but five to nine alleles were found. In studies in the Philippines (SAKAMURI et al., 2009), Thailand, Indonesia, Korea and Japan (ZHANG; BUDIAWAN; MATSUOKA, 2005), AC8b was found to be more frequent with eight repeats.

In our study, AC9 also showed greater allelic diversity ($h = 0.51$) than the study by Fontes et al. (2009) in Brazil ($h = 0.28$ on average). As for the number of alleles, in our study seven to nine alleles were observed, with seven and eight being the most frequent. Studies in other countries (SAKAMURI et al., 2009; SRISUNGNAM et al., 2009; ZHANG; BUDIAWAN; MATSUOKA, 2005), even in a few samples, found 10 alleles of this locus. This shows that 10 repeats of AC9 is quite rare and has not yet been published in samples from Brazil.

Thus, despite the small number of samples analyzed in this study, the AC9 and AC8b *loci* showed greater diversity in the samples from Fortaleza than Sao Paulo and Rio Janeiro (FONTES et al., 2009), but with a similar number of alleles.

In this study, the AC8a *locus* was moderately variable ($h= 0.57$), with eight to 11 alleles, but in several published studies (FONTES et al., 2009; SAKAMURI et al., 2009; SRISUNGNAM et al., 2009; WENG et al., 2007; ZHANG; BUDIAWAN; MATSUOKA, 2005), this *locus* showed more variability in terms of the number of alleles (6 to 13).

As for the GTA9 *locus*, in our study it was also moderately variable ($h = 0.57$), with eight to 16 alleles, but in other published studies (FONTES et al., 2009; SAKAMURI et al., 2009; SRISUNGNAM et al., 2009), this locus showed more variability in terms of the number of alleles (7 to 15) and in the study by Weng et al (2007) in China, this locus had nine to 45 alleles ($h = 0.82$). In addition, in these studies this locus had nine alleles with the highest frequency, while in our study the most frequent number of repeats was 10.

The great genetic diversity of the isolates in Brazil, already reported by Fontes et al. (2009), is demonstrated in this study by the formation of four groups with only 12 samples and by the fact that the strains with unique genotypes are geographically closer than those with similar genotypes. This genetic diversity is similar in number to that found in India (SHINDE et al., 2009) and may be related to the fact that these two countries alone are responsible for 69% of new leprosy cases detected in the world (WHO, 2013). Thus, this great genetic variability of *M. leprae* would be proportional to the number of leprosy cases in these countries.

In the current study, patients from the same municipality were used, so due to the small distance between them, variations in a few VNTR *loci* are quite significant for the formation of groups as demonstrated by Matsuoka et al. (2004) who applied the microsatellite *locus (*TTC) 21 to genotype *M. leprae* obtained from a father and son who lived in the same household, publishing that there had been infections of different bacilli with 10 and 18 copies, respectively. In addition, genotypes of TTC repeats were found to differ between a son undergoing treatment and two brothers. These results showed the possibility that, in addition to exposure through the presence of a patient with multibacillary leprosy who lived with family members, there may have been other sources of infection to which the patient's household members had been exposed outside the home (MATSUOKA et al., 2004b).

The formation of groups using four *loci* may reflect the formation of groups with more *loci*, as suggested in a study with *M. tuberculosis*, in which Warren et al. (2002) associated a specific region of the city with the formation of groups and this association was maintained even when some genetic differences were tolerated. This also suggests that the rigorous definition of genetic groups may underestimate the levels of clustering within a specific population, making it difficult to understand the chains of transmission of a disease (WARREN et al., 2002).

Other studies have been published studying the molecular epidemiology of leprosy with few VNTR loci, such as Truman et al. (2004) and Da Silva Rocha et al. (2011). These authors used 4 VNTRs (GAA, GTA9, AT17 and TA18) to differentiate unrelated strains of *M. leprae* and Phetsuksiri et al. (2012) used *rpot* and TTC genotyping to study the transmission and distribution of leprosy in Thailand.

When forming groups with strains with identical genotypes, due to the loss of demographic data, only 24 samples out of 48 were used for descriptive analysis. It was not possible to establish the exact contact between the individuals belonging to the groups, but some similarities were observed between them.

Thus, the probable association between the individuals in group one was that they were both born in a municipality in the interior of Ceará called Aracati. This is a small town with a population density of 61.25 inhabitants/km^2 (GOVERNO MUNICIPAL DE ARACATI, 2014), in which the individuals lived in the same period for 12 years and may have had some kind of contact. In group two, it was observed that the two individuals belonged to the same age group, were diagnosed in the same year and had the same clinical form of the disease. Group three was also made up of individuals from the same age group, but with a year's difference in diagnosis. Group four was the group with the largest number of individuals and all were diagnosed in the same year, with the two virchowian cases diagnosed in the same month.

The most frequent genotypes in the current study were AC8b: 8, AC9: 7, AC8a: 8, GTA9: 10 and AC8b: 7, AC9: 8, AC8a: 9, GTA9: 9. The former was rarely found in other studies (KIMURA et al., 2009; SAKAMURI et al., 2009; SRISUNGNAM et al., 2009), including in samples from Brazil (FONTES et al., 2009), which may represent a genotype of strains circulating more frequently in Ceará.

When comparing the samples from the groups with those that did not form groups, it was observed that patients who were older (mean age 45 years) had strains that formed groups because they had similar genotypes, while patients who were exposed early (mean age 25 years, $p = 0.0110$) had strains that had unique genotypes. From this it can be assumed that transmission among older patients may have been more recent and with slightly earlier symptoms. As a result, it was observed that the time between symptoms and diagnosis for samples with similar genotypes was slightly shorter than for those with a single genotype. From this it can be conjectured that the single genotype strains may have had more time to mutate and acquire greater genetic variability in the host.

Although the limited number of samples studied in the current study prevents further grouping and statistical analysis with greater significance, the allelic diversity between some *loci in the* DNA of the patients' bacilli suggests the existence of different strains of *M. leprae* circulating in the municipality under study, since *M. leprae* undergoes few mutations (CLARK-CURTISS; WALSH, 1989; COLE et al., 2001). This demonstrates that the continued incidence of leprosy in this community was the result of a complex series of transmission events, rather than the recent transmission of one or a few dominant stable clones.

In the present study, a dispersion of similar genotypes was found, since no groupings of patients with similar genotypes were observed in terms of geographical distribution. This can be seen when comparing studies in various countries such as Colombia (CARDONA-CASTRO et al., 2009), the Philippines (SAKAMURI et al., 2009), India (SHINDE et al., 2009), Thailand (SRISUNGNAM et al., 2009), China (XING et al., 2009), with the current study and another study carried out in Brazil (FONTES et al., 2009) and it can be seen how much more dispersed the Brazilian genotypes are than the others.

Some studies have analyzed the genetic diversity of *M. leprae* by comparing samples collected at the same time from different sites on the patient's body (MONOT et al., 2008; SAKAMURI et al., 2009; XING et al., 2009; YOUNG et al., 2004) or over different periods of time, as in the study by Lavania et al (2011). In this study, different strains were demonstrated in a patient when he was first diagnosed in 2004 and in the repeated examination in 2007, in which the patient had already been discharged due to the completion of therapy. The bacillus isolated in 2007 from the biopsy revealed a different number of STR repeats indicating infection with a new strain circulating in the community (LAVANIA et al., 2011).

The present study analyzed SN and BP samples collected at the time of the individual's diagnosis and when comparing the samples, notable differences were observed in two individuals with strains that differed at seven *loci* and in four individuals with differences at five *loci*. This genetic diversity of *M. leprae* found in the individual may represent a mixed infection in the patient, who may be carrying two subpopulations of bacteria that have different dominant genotypes and have presumably expanded independently.

This detection of different strains in the same individual may also be due to the unique sequence revealed by PCR amplification in each sample corresponding to the dominant genotype within a heterogeneous population. Entry into the host provides an opportunity for a limited number of bacteria to undergo clonal expansion in a new, relatively isolated niche. This situation could allow an initially subdominant population to emerge, either through a purely random process or as a result of some selective advantage. The intergenic or interpseudogenic location of VNTR *loci* makes it unlikely that the difference in copy number repetition itself confers any biological advantage, but it is possible that it acts as a marker for some biologically distinct subpopulations of bacteria and is passively carried during the selective expansion of that subpopulation (YOUNG et al., 2004).

In the current study, the greatest similarities between SN and BP samples from the same individual were observed in seven patients. Of these, five did not differ in any amplified *loci,* but the SN of two patients did not amplify a locus and the SN of three patients did not amplify two *loci*; and two patients differed in two *loci,* but amplified all *loci*. The number of individuals with similar strains was small when compared to the result found by Monot et al (2008), who evaluated 4 VNTR *loci* in a comparative analysis of biopsies from different sites of the same patient and from the same site at two different times and revealed no variation in 36 out of 42 patients. Of these, four patients had their two strains varying at a single *locus*, one patient at three *loci*, and the other at all four *loci* investigated. The greatest variability was seen at the 21_TTC locus, followed by 17_AT. Most of the differences were due to a loss or gain of a single repeat unit, with the two strains from one patient showing differences of two and three units at the 14_AT and 9_GTA loci, respectively.

However, Monot et al. (2008) only compared biopsy samples, while in our study BP and SN samples were compared and because they are different sites, the bacilli may undergo different adaptations to the pressures exerted by the host's immune response and to the physiological conditions of the different anatomical sites. The bacillus in the SN sample represents the strain at the host's point of entry that may be circulating in the community and the bacillus in the BP sample represents the strain that has already established itself and may have undergone clonal expansion and genotypic differentiation.

Most of the samples from the same patient with different VNTRs identified in this study involved an increase in the number of copies in the BP samples compared to the SN. This was also observed in the study by Young et al. (2004), but this only analyzed 7 patients, while in our study 38 patients were studied. The increase in the size of the microsatellites between samples from the same individual in the study by Young et al. (2004) was only observed when comparing skin biopsy samples and nerve biopsies from seven of the eight patients, with the highest number of copies in the VNTR loci of the strains from the nerve biopsy samples. When in this study the authors compared SN samples with BP samples, earlobe lymph and peripheral blood mononuclear cells from seven patients with lepromatous leprosy, there was an exact match in the number of copies of each of the samples from the same patient, although they were unable to obtain results for all the loci from all the sites, especially the SN, as in our study.

The higher number of copies in the BP samples may be related to some characteristic of *M. leprae* replication in the skin, such as an accelerated duplication time, promoting a change in the number of copies of the VNTR *locus* in a way that results in an alteration of the overall dominant genotype (YOUNG et al., 2004). In addition, the increase in the size of the microsatellites may be associated with an increase in the pathogenicity of the strain, since if there is more than one type of strain circulating in the community, the strain from the BP sample would be more pathogenic than the one from the SN sample, since the SN strain will not necessarily invade the host and cause the disease, while the BP strain has already invaded the host and caused widespread disease. These results are reminiscent of the increased size of microsatellites being associated with disease in humans (CUMMINGS; ZOGHBI, 2000; KENNEDY et al., 2003).

Whatever the underlying mechanism, the finding of differences in *M. leprae* subpopulations between BP and SN samples highlights the importance of better understanding the factors involved in tissue invasion by *M. leprae.*

In the study by Lavania et al. (2011), *loci* 12-5, 21-3, 18-8 and 27-5 were polymorphic. In China, they showed no variation (WENG et al., 2007). This shows that *loci* that are polymorphic may present themselves in different ways, which makes it necessary to carry out studies with VNTRs to complement the database on the genotypes of *M. leprae* prevalent in various areas of the world and to understand whether the behavior observed is exclusively associated with the region of study or the number of samples that were evaluated.

Further studies of genetic diversity in samples with known epidemiological links will be important to assess the extent to which microsatellites can be used as reliable markers of transmission events.

This study has shown that it is possible to use SN samples, which are less invasive and easier to collect, for a molecular epidemiological approach to the study of leprosy. However, in order to better discriminate between strains, more loci must be included in the typing scheme, which will depend on the type of sample collected and the patient's clinical form of leprosy. Furthermore, in order to continue exploring the diversity of microsatellites, it is important to discover other forms of genetic variation suitable for typing. Important objectives will be to identify typing systems capable of providing reliable information on the transmission dynamics of *M. leprae* and to use these to help find interventions to reduce the number of new leprosy cases.

7 Conclusion

- *M. leprae* DNA extracted from nasal secretions was amplified in 69.2% of cases, 66.9% of the Co group and 28.1% of the GE group.

- The detection of *M. leprae* DNA in nasal secretions was influenced by age, sex, baciloscopic index and socioeconomic status.

- The fact that the individual was male increased the chance of him being PCR positive by 6.266 times.

- Each year that the individual got older implied a 1.046 times greater chance of them presenting a positive PCR for RLEP.

- The fact that the individual belonged to socio-economic class D/E increased the chance of them being PCR positive by 3.083 times.

- In the analysis of the geographic distribution of the groups' homes, the RLEP-positive individuals in the Co group were distributed throughout the municipality and were the intersection group between the RLEP-positive individuals in the C and GE groups.

- In the analysis of the geographical distribution of housing in group C, patients with positive bacilloscopy and positive RLEP were more clustered than patients with positive bacilloscopy and negative RLEP.

- The molecular technique with amplification of the RLEP region of *M. leprae* can be used to monitor cases and controls, identifying the population at risk of developing leprosy and individuals without clinical signs of the disease, but who may be playing an active role in the transmission of leprosy.

- The most appropriate genetic markers for differentiating *M. leprae* strains in nasal secretion samples were the four VNTR loci: AC8b, AC9, AC8a and

GTA9; which showed an allelic diversity considered to be moderately discriminating.

- Four groups with identical *M. leprae* genotype strains were formed, with a Hunter - Gaston discriminatory index (HGDI) of 0.7273.

- The most frequent genotypes were AC8b: 8, AC9: 7, AC8a: 8, GTA9: 10 and AC8b: 7, AC9: 8, AC8a: 9, GTA9: 9.

- Patients who were older had strains that were similar, while patients who were exposed early had strains with unique genotypes.

- The time between symptoms and diagnosis for the grouped samples was shorter than for the single genotype samples.

- The allelic diversity between the *loci of the* bacilli DNA among the patients suggested the existence of different strains of *M. leprae* circulating in the municipality under study.
- In the analysis of the geographical distribution of housing, the dispersion of similar genotypes was observed.
- Differences in *M. leprae* subpopulations were found between BP and SN samples from the same individual.
- It is possible to use SN samples, which are less invasive and easier to collect, for a molecular epidemiological approach to the study of leprosy.
- In NS samples, the limited discriminatory capacity of 4 *loci* in epidemiological analysis implies the need to look for other useful polymorphic *loci* for detailed subdivision of clinical isolates.

REFERENCES

ALLIX-BÉGUEC, C. et al. Evaluation and strategy for use of MIRU-VNTRplus, a multifunctional database for online analysis of genotyping data and phylogenetic identification of Mycobacterium tuberculosis complex isolates. **Journal of clinical microbiology**, v. 46, n. 8, p. 2692-2699, 2008.

ALMEIDA, E. C. DE et al. Detection of Mycobacterium leprae DNA by polymerase chain reaction in the blood and nasal secretion of Brazilian household contacts. **Memorias do Instituto Oswaldo Cruz**, v. 99, n. 5, p. 509-511, 2004.

ANTONIO, L.; SANTOS, D. C.; ARMAUER, H. Counterpoints in the history of leprosy in Brazil: scenarios of stigma and confinement. **Revista Brasileira de Estudos de Populaçâo**, v. 21, p. 167-190, 2008.

ARMOUR, K. S.; SCOLYER, R. A.; BARNETSON, R. S. Borderline lepromatous leprosy presenting as a single cutaneous plaque.**The Australasian journal of dermatology**, 2005.

BABU, M. M. Did the loss of sigma factors initiate pseudogene accumulation in M. leprae?**Trends in Microbiology**, 2003.

BAKKER, M. I. et al. Risk factors for developing leprosy--a population-based cohort study in Indonesia. **Leprosy review**, v. 77, n. 1, p. 48-61, 2006.

BANERJEE, S. et al. Multiplex PCR technique could be an alternative approach for early detection of leprosy among close contacts--a pilot study from India. **BMC infectious diseases**, v. 10, p. 252, 2010.

BARKER, L. P. Mycobacterium leprae interactions with the host cell: recent advances. **The Indian journal of medical research**, v. 123, n. 6, p. 748-59, jun. 2006.

BEIGUELMAN, B. Genètica e hanseniase * Genetics and leprosy. **Ciencia & Saùde Coletiva**, v. 7, n. 1, p. 117-128, 2002.

BEYENE, D. et al. Nasal carriage of Mycobacterium leprae DNA in healthy individuals in Lega Robi village, Ethiopia. **Epidemiology and infection**, v. 131, n. 2, p. 841-848, 2003.

BRITTON, W. J.; LOCKWOOD, D. N. J. **LeprosyLancet. Proceedings**...2004

BROWN, R. M. et al. Lipoarabinomannan-reactive human secretory immunoglobulin A responses induced by mucosal bacille Calmette-Guérin vaccination. **The Journal of infectious diseases**, v. 187, n. 3, p.513-517, 2003.

BUHRER-SEKULA, S. et al. Use of ML dipstick as a tool to classify leprosy patients. **International journal ofleprosy and other mycobacterial diseases : official organ of the International Leprosy Association**, v. 68, n. 4, p. 456-463, 2000.

CARDONA-CASTRO, N. et al. Identification and comparison of Mycobacterium leprae genotypes in two geographical regions of Colombia. **Leprosy review**, v. 80, n. 3, p. 316-21, set. 2009.

CARDONA-CASTRO, N.; BELTRAN-ALZATE, J. C.; MANRIQUE-HERNANDEZ, R. Survey to identify

Mycobacterium leprae-infected household contacts of patients from prevalent regions of leprosy in Colombia. **Memórias do Instituto Oswaldo Cruz**, v. 103, n. 4, p. 332-6, jun. 2008.

CHEN, L. et al. Single intranasal mucosal Mycobacterium bovis BCG vaccination confers improved protection compared to subcutaneous vaccination against pulmonary tuberculosis. **Infection and immunity**, v. 72, n. 1, p. 238-246, 2004.

CLARK-CURTISS, J. E.; WALSH, G. P. isolates of Mycobacterium leprae . Conservation of Genomic Sequences among Isolates of MVycobacterium leprae. **Journal of Bacteriology**, v. 171, n. 9, 1989.

COLE, S. T. et al. Massive gene decay in the leprosy bacillus. **Nature**, v. 409, n. 6823, p. 1007-1011, 2001.

CONVIT, J. et al. Immunoprophylactic trial with combined Mycobacterium leprae/BCG vaccine against leprosy: preliminary results. **Lancet**, v. 339, n. 8791, p. 446-50, 1992.

CROFT, R. A.; CROFT, R. P. Expenditure and loss of income incurred by tuberculosis patients before reaching effective treatment in Bangladesh. **The international journal of tuberculosis and lung disease : the official journal of the International Union against Tuberculosis and Lung Disease**, v. 2, n. 3, p. 252-254, 1998.

CUEVAS, J. et al. Erythema Nodosum Leprosum: Reactional Leprosy. **Seminars in Cutaneous Medicine and Surgery**, v. 26, n. 2, p. 126-130, 2007.

CUMMINGS, C. J.; ZOGHBI, H. Y. Fourteen and counting: unraveling trinucleotide repeat diseases. **Human molecular genetics**, v. 9, n. 6, p. 909-16, 12 abr. 2000.

CUNHA, S. S. et al. Design of the leprosy component of the Brazilian BCG revaccination trial for assessing BCG effectiveness against leprosy in school children. **International journal of leprosy and other mycobacterial diseases : official organ of the International Leprosy Association**, v. 72, n. 1, p. 8-15.

CURY, M. R. DE C. O. et al. Spatial analysis of leprosy incidence and associated socioeconomic factors. **Revista de Saùde Pùblica**, v. 46, n. 1, p. 110-118, 2012.

CUSTODIO, L. A. et al. Detection of Lsr2 Gene of Mycobacterium leprae in Nasal Mucus. **Brazilian archives of biology and technology**, v. 55, n. June, p. 375-379, 2012.

DA CUNHA, M. D. et al. The impact of leprosy elimination strategy on an endemic municipality in Ric de Janeiro State, Brazil. **Cadernos de saude publica / Ministerio da Saude, Fundacao Oswaldo Cruz, Escola Nacional de Saude Publica**, v. 23, n. 5, p. 1187-1197, 2007.

DA SILVA ROCHA, A. et al. Genotyping of Mycobacterium leprae from Brazilian leprosy patients suggests the occurrence of reinfection or of bacterial population shift during disease relapse. **Journal of Medical Microbiology**, v. 60, n. 10, p, 1441-1446, 2011.

DATASUS. Available at: < www.datasus.gov.br>. Accessed on: December 20, 2013.

DE WIT, M. Y. et al. Application of a polymerase chain reaction for the detection of Mycobacterium leprae in skin tissues. **Journal of clinical microbiology**, v. 29, n. 5, p. 906-910, 1991a.

DE WIT, M. Y. et al. Application of a polymerase chain reaction for the detection of Mycobacterium leprae in skin tissues. **Journal of clinical microbiology**, v. 29, n. 5, p. 906-10, May 1991b.

DONOGHUE, H. D.; HOLTON, J.; SPIGELMAN, M. PCR primers that can detect low levels of Mycobacterium leprae DNA. **Journal of medical microbiology**, v. 50, n. 2, p. 177-82, feb. 2001.

DOUGLAS, J. T. et al. Prospective study of serological conversion as a risk factor for development of leprosy among household contacts. **Clinical and diagnostic laboratory immunology**, v. 11, n. 5, p. 897-900, 2004.

DÜPPRE, N. C. et al. Effectiveness of BCG vaccination among leprosy contacts: a cohort study. **Transactions of the Royal Society of Tropical Medicine and Hygiene**, v. 102, n. 7, p. 631-638, 2008.

FINE, P. E. M. et al. Protective efficacy of BCG against leprosy in Northern Malawi. **Lancet**, v. 2, n. 8505, p.499-502, 1986.

FONTES, A. N. B. et al. Genetic diversity of mycobacterium leprae isolates from Brazilian leprosy patients. **Leprosy review**, v. 80, n. 3, p. 302-15, 2009.

FOSS, N. T.; CAROLINA, A.; MOTTA, F. Leprosy , a neglected disease that causes a wide variety of clinical conditions in tropical countries. **Memórias do Instituto Oswaldo Cruz**, v. 107, p. 28-33, 2012.

FOSS, N. T.; MOTTA, A. C. F. Leprosy, a neglected disease that causes a wide variety of clinical conditions in tropical countries. **Memórias do Instituto Oswaldo Cruz**, v. 107 Suppl , p. 28-33, 2012.

FROTA, C. C. et al. Seropositivity to anti-phenolic glycolipid-I in leprosy cases, contacts and no known contacts of leprosy in an endemic and a non-endemic area in northeast Brazil. **Transactions of the Royal Society of Tropical Medicine and Hygiene**, v. 104, n. 7, p. 490-495, 2010a.

FROTA, C. C. et al. Detection of Mycobacterium leprae in water reservoirs in regions with high and low leprosy prevalence in the State of Cearà, Northeast Brazil. **Revista Eletrônica Pesquisa Médica**, v. 4, n. 1, p. 1-7, 2010b.

FROTA, C. C. et al. Mycobacterium leprae in six-banded (Euphractus sexcinctus) and nine-banded armadillos (Dasypus novemcinctus) in Northeast Brazil. **Memórias do Instituto Oswaldo Cruz**, v. 107 Suppl , p.209-13, dec. 2012.

GILLIS, T. et al. Characterization of short tandem repeats for genotyping Mycobacterium leprae. **Leprosy review**, v. 80, n. 3, p. 250-260, 2009.

GILMOUR, M. N.; WHITTAM, T. S. Methods of Multilocus Enzyme Electrophoresis for Bacterial Population Genetics and Systematics. **Applied and Environmental Microbiology**, v. 51, n. 5, p. 873884, 1986.

GLATMAN-FREEDMAN, A. The role of antibody-mediated immunity in defense against Mycobacterium tuberculosis: Advances toward a novel vaccine strategy. **Tuberculosis**, v. 86, n. 3-4 SPEC. ISS., p. 191-197, 2006.

GOONETILLEKE, N. P. et al. Enhanced immunogenicity and protective efficacy against Mycobacterium tuberculosis of bacille Calmette-Guérin vaccine using mucosal administration and boosting with a recombinant modified vaccinia virus Ankara. **Journal of immunology (Baltimore, Md. : 1950)**, v. 171, n. 3, p. 1602-1609, 2003.

GOULART, L. R.; GOULART, I. M. B. Leprosy pathogenetic background: a review and lessons from other mycobacterial diseases. **Archives for dermatological research. Archiv fur dermatologische Forschung**, v. 301, n. 2, p. 123-137, 2009.

MUNICIPAL GOVERNMENT OF ARACATI. **Aracati**. Available at: <http://www.aracati.ce.gov.br/>. Accessed on: February 24, 2014.

GROATHOUSE, N. A. et al. Multiple polymorphic loci for molecular typing of strains of Mycobacterium leprae. **Journal of clinical microbiology**, v. 42, n. 4, p. 1666-1672, 2004.

HAUSFELD, R. G. An anthropological method for measuring exposure to leprosy in a leproxy- endemic population at Karimui, New Guinea. **Bulletin of the World Health Organization**, v. 43, n. 6, p. 863-877, 1970.

HIROI, T. et al. Nasal immune system: distinctive Th0 and Th1/Th2 type environments in murine nasal-associated lymphoid tissues and nasal passage, respectively. **European journal of immunology**, v. 28, n. 10, p. 3346-3353, 1998.

HUNTER, P. R.; GASTON, M. A. Numerical index of the discriminatory ability of typing systems: an application of Simpson's index of diversity. **Journal of clinical microbiology**, v. 26, n. 11, p. 24652467, 1988.

JENSEN, R. W. et al. DNA fingerprinting of Mycobacterium leprae strains using variable number tandem repeat (VNTR) - fragment length analysis (FLA). **Journal of visualized experiments**, n. 53, p. e3104, 2011.

JOB, C. K. et al. Transmission of leprosy: a study of skin and nasal secretions of household contacts of leprosy patients using PCR. **The American journal of tropical medicine and hygiene**, v. 78, n. 3, p. 518-21, mar. 2008.

JOYCE, M. P. Historic aspects of human susceptibility to leprosy and the risk of conjugal transmission. **Memórias do Instituto Oswaldo Cruz**, v. 107, p. 17-21, 2012.

KÀLLENIUS, G. et al. Should a new **tuberculosis** vaccine be administered intranasally? **Tuberculosis (Edinburgh, Scotland)**, v. 87, n. 4, p. 257-66, jul. 2007.

KEIM, P. et al. Multiple-locus variable-number tandem repeat analysis reveals genetic relationships within Bacillus anthracis. **Journal of bacteriology**, v. 182, n. 10, p. 2928-36, May 2000.

KENNEDY, L. et al. Dramatic tissue-specific mutation length increases are an early molecular event in Huntington disease pathogenesis. **Human molecular genetics**, v. 12, n. 24, p. 3359-67, Dec. 15, 2003.

KERR-PONTES, L. R. S. et al. Inequality and leprosy in Northeast Brazil: an ecological study.

International journal of epidemiology, v. 33, n. 2, p. 262-269, 2004.

KERR-PONTES, L. R. S. et al. Socioeconomic, environmental, and behavioral risk factors for leprosy in North-east Brazil: results of a case-control study. **International journal of epidemiology**, v. 35, n. 4, p. 994-1000, 2006.

KIMURA, M. et al. Rapid variable-number tandem-repeat genotyping for Mycobacterium leprae clinical specimens. **Journal of clinical microbiology**, v. 47, n. 6, p. 1757-66, jun. 2009.

KLATSER, P. R. et al. Detection of Mycobacterium leprae nasal carriers in populations for which leprosy is endemic. **Journal of clinical microbiology**, v. 31, n. 11, p. 2947-51, nov. 1993.

KUMAR, B.; DOGRA, S. Leprosy: A disease with diagnostic and management challenges! **Indian Journal of Dermatology, Venereology and Leprology**, v. 75, n. 2, p. 111, 2009.

LAVANIA, M. et al. Detection of viable Mycobacterium leprae in soil samples: insights into possible sources of transmission of leprosy. **Infection, genetics and evolution : journal of molecular epidemiology and evolutionary genetics in infectious diseases**, v. 8, n. 5, p. 627-31, sep. 2008.

LAVANIA, M. et al. Molecular typing of Mycobacterium leprae strains from northern India using short tandem repeats. **The Indian journal of medical research**, v. 133, n. June, p. 618-26, jun. 2011.

LIU, D. et al. Analysis of newly detected leprosy cases and misdiagnosis in Wuhan (1990-2004). **Leprosy review**, v. 80, n. 4, p. 410-415, 2009.

LOBATO, J. et al. Comparison of three immunological tests for leprosy diagnosis and detection of subclinical infection. **Leprosy review**, v. 82, n. 4, p. 389-401, 2011.

LOCKWOOD, D. N. J.; SUNEETHA, S. Leprosy: too complex a disease for a simple elimination paradigm. **Bulletin of the World Health Organization**, v. 83, n. 3, p. 230-235, 2005.

LOMBARDI, C. et al. Protective efficacy of BCG against leprosy in Sao Paulo. **Bull Pan Am Health Organ**, v. 30, n. 1, p. 24-30, 1996.

LOVETT, S. T. Encoded errors: mutations and rearrangements mediated by misalignment at repetitive DNA sequences. **Molecular microbiology**, v. 52, n. 5, p. 1243-53, 2004.

MALORNY, B.; JUNKER, E.; HELMUTH, R. Multi-locus variable-number tandem repeat analysis for outbreak studies of Salmonella enterica serotype Enteritidis. **BMC microbiology**, v. 8, p. 84, 2008.

MATSUOKA M et al. Mycobacterium leprae DNA in daily using water as a possible source of leprosy infection. **Indian Journal of Leprosy**, v. 71, n. 1, p. 61-67, 1999.

MATSUOKA, M. et al. Genotyping of Mycobacterium leprae on the Basis of the Polymorphism of TTC Repeats for Analysis of Leprosy Transmission. **Journal of Clinical Microbiology**, v. 42, n. 2, p. 741-745. 2004a.

MATSUOKA, M. et al. Genotyping of Mycobacterium leprae on the basis of the polymorphism of TTC repeats for analysis of leprosy transmission. **Journal of clinical microbiology**, v. 42, n. 2, p. 741-745, 2004b.

MATSUOKA, M.; KASHIWABARA, Y.; NAMISATO, M. A Mycobacterium leprae isolate resistant to dapsone, rifampin, ofloxacin and sparfloxacin. **International journal of leprosy and other mycobacterial diseases : official organ of the International Leprosy Association**, v. 68, n. 4, p. 452-5, dec. 2000.

MEYERS, W. M.; GORMUS, B. J.; WALSH, G. P. Nonhuman sources of leprosy. **International journal ofleprosy and other mycobacterial diseases : official organ of the International Leprosy Association**, v. 60, n. 3, p. 477-480, 1992.

MINISTRY OF HEALTH. Available at: < www.saude.gov.br>. Accessed on: December 20, 2013.

MONOT, M. et al. On the origin of leprosy. **Science (New York, N.Y.)**, v. 308, n. 5724, p. 10401042, 2005.

MONOT, M. et al. Are variable-number tandem repeats appropriate for genotyping Mycobacterium leprae? **Journal of clinical microbiology**, v. 46, n. 7, p. 2291-7, jul. 2008.

MONOT, M. et al. Comparative genomic and phylogeographic analysis of Mycobacterium leprae. **Nature genetics**, v. 41, n. 12, p. 1282-9, dec. 2009.

NAGAO-DIAS, A. T. et al. Salivary anti-PGL IgM and IgA titers and serum antibody IgG titers and avidities in leprosy patients and their correlation with time of infection and antigen exposure. **The Brazilian journal of infectious diseases : an official publication of the Brazilian Society of Infectious Diseases**, v. 11, n. 2, p. 215-219, 2007.

NAKAMURA, K. et al. Detection of RNA expression from pseudogenes and non-coding genomic regions of Mycobacterium leprae. **Microbial Pathogenesis**, v. 47, n. 3, p. 183-187, 2009.

PARDILLO, F. E. F. et al. Methods for the classification of leprosy for treatment purposes. **Clinical infectious diseases : an official publication of the Infectious Diseases Society of America**, v. 44, n. 8, p. 1096-1099, 2007.

PATROCÌNIO, L. G. et al. Detection of Mycobacterium leprae in nasal mucosa biopsies by the polymerase chain reaction. **FEMS Immunology and Medical Microbiology**, v. 44, n. 3, p. 311-316, 2005.

PHETSUKSIRI, B. et al. SNP genotypes of Mycobacterium leprae isolates in Thailand and their combination with rpoT and TTC genotyping for analysis of leprosy distribution and transmission. **Japanese journal of infectious diseases**, v. 65, n. 1, p. 52-6, 2012.

PLIKAYTIS, B. B.; GELBER, R. H.; SHINNICK, T. M. Rapid and sensitive detection of Mycobacterium leprae using a nested-primer gene amplification assay. **Journal of clinical microbiology**, v. 28, n. 9, p. 1913-1917, 1990.

R, M. C. et al. Hansen's disease: Review of a case. **Revista chilena de infectologia**, v. 25, n. 1, p. 64-69, 2008.

RABELLO, F. E. A clinico-epidemiological classification of the forms of leprosy. **Int. J. Leprosy,**

n. 5, p. 343-356, 1937.

RADA, E. et al. Serologic follow-up of IgG responses against recombinant mycobacterial proteins ML0405, ML2331 and LID-1 in a leprosy hyperendemic area in Venezuela. **Mem Inst Oswaldo Cruz**, v. 107 Suppl , p. 90-94, 2012.

REES, R. J. Animal models in leprosy. **British medical bulletin**, v. 44, n. 3, p. 650-64, jul. 1988.

REYES, J. F.; TANAKA, M. M. Mutation rates of spoligotypes and variable number tandem repeat loci in Mycobacterium tuberculosis. **Infection, genetics and evolution : journal of molecular epidemiology and evolutionary genetics in infectious diseases**, v. 10, n. 7, p. 1046-1051, 2010.

RIDLEY, D. S.; JOPLING, W. H. Classification of leprosy according to immunity. A five-group system. **International journal of leprosy and other mycobacterial diseases.** v. 34, n. 3, p. 255-273, 1966.

RODRIGUES, L. C. et al. Long lasting BCG protection against leprosy. **Vaccine**, v. 25, n. 39-40, p. 6842-6844, 2007.

RODRIGUES, M. L. et al. Protective effect of intradermal BCG against leprosy; a case-control study in central Brazil. **International journal of leprosy and other mycobacterial diseases**, v. 62, n. 4, p. 532-538, 1994.

SAKAMURI, R. M. et al. Population-based molecular epidemiology of leprosy in Cebu, Philippines. **Journal of clinical microbiology**, v. 47, n. 9, p. 2844-54, sep. 2009.

SALES, A. M. et al. Leprosy among patient contacts: A multilevel study of risk factors. **PLoS**

Neglected Tropical Diseases, v. 5, n. 3, 2011.

SASAKI, S. et al. Mycobacterium leprae and leprosy: a compendium. **Microbiology and immunology**, v. 45, n. 11, p. 729-736, 2001.

SCOLLARD, D. M. et al. The continuing challenges of leprosy. **Clinical microbiology reviews**, v. 19, n. 2, p. 338-381, 2006.

SECRETARIAT OF HEALTH OF THE GOVERNMENT OF THE STATE OF CEARA. Leprosy Epidemiological Report. Available at: <www.saude.ce.gov.br/index.php/boletins>. Accessed on: December 20, 2013.

SHEPARD, C. C. THE EXPERIMENTAL DISEASE THAT FOLLOWS THE INJECTION OF HUMAN LEPROSY BACILLI INTO FOOT-PADS OF MICE. **The Journal of experimental medicine**, v. 112, n. 3, p. 445-454, 1960.

SIMPSON, E. H. Measurement of diversity. **Nature,** p. 163-688, 1949.

SHIN, Y. C. et al. Variable numbers of TTC repeats in Mycobacterium leprae DNA from leprosy patients and use in strain differentiation. **Journal of clinical microbiology**, v. 38, n. 12, p. 4535-4538, 2000.

SHINDE, V. et al. VNTR typing of Mycobacterium leprae in South Indian leprosy patients. **Leprosy review**, v. 80, n. 3, p. 290-301, sep. 2009.

SINGH, P.; COLE, S. T. Mycobacterium leprae: genes, pseudogenes and genetic diversity. **Future microbiology**, v. 6, n. 1, p. 57-71, 2011.

SOLA, C. et al. Genotyping of the Mycobacterium tuberculosis complex using MIRUs: association with VNTR and spoligotyping for molecular epidemiology and evolutionary genetics. **Infection, Genetics and Evolution**, v. 3, n. 2, p. 125-133, jul. 2003.

SRISUNGNAM, S. et al. Molecular epidemiology of leprosy based on VNTR typing in Thailand.

Leprosy review, v. 80, n. 3, p. 280-9, set. 2009.

SUPPLY, P. et al. Proposal for standardization of optimized mycobacterial interspersed repetitive unit-variable-number tandem repeat typing of Mycobacterium tuberculosis. **Journal of clinical microbiology**, v. 44, n. 12, p. 4498-4510, 2006.

TALHARI, S. et al. Hansen's disease: a vanishing disease? **Memórias do Instituto Oswaldo Cruz**, v. 107, n.5, p. 13-16, 2012.

THIBAULT, V. C. et al. Combined multilocus short-sequence-repeat and mycobacterial interspersed repetitive unit-variable-number tandem-repeat typing of Mycobacterium avium subsp.

paratuberculosis isolates. **Journal of clinical microbiology**, v. 46, n. 12, p. 4091-4094, 2008.

TRUMAN, R. et al. Genotypic Variation and Stability of Four Variable-Number Tandem Repeats and Their Suitability for Discriminating Strains of Mycobacterium leprae. **Journal of Clinical Microbiology**. v. 42, n. 6, p. 2558-2565, 2004.

TRUMAN, R. Leprosy in wild armadillos. **Leprosy review**, v. 76, n. 3, p. 198-208, 2005.

TRUMAN, R. W. et al. Evaluation of the origin of Mycobacterium leprae infections in the wild armadillo, Dasypus novemcinctus. **The American journal of tropical medicine and hygiene**, v. 35, n. 3, p. 588-593, 1986.

TYAGI, J. S.; SAINI, D. K. Did the loss of two-component systems initiate pseudogene accumulation in Mycobacterium leprae? **Microbiology (Reading, England)**, v. 150, n. Pt 1, p. 4-7, jan. 2004.

VISSA, V. D.; BRENNAN, P. J. The genome of Mycobacterium leprae: a minimal mycobacterial gene set. **Genome biology**, v. 2, n. 8, p. REVIEWS1023, 2001.

WALKER, S. L.; LOCKWOOD, D. N. J. The clinical and immunological features of leprosy. **British medical bulletin**, v. 77-78, p. 103-121, 2006.

WARREN, R. M. et al. Evolution of the IS 6110- Based Restriction Fragment Length Polymorphism Pattern during the Transmission of Mycobacterium tuberculosis. **Journal of clinical Microbiology**, v. 40, n. 4, p. 1277-1282 , 2002.

WENG, X. et al. Identification and Distribution of Mycobacterium leprae Genotypes in a Region of High Leprosy Prevalence in China : a 3-Year Molecular Epidemiological Study. **Journal of clinical Microbiology**, v. 45, n. 6, p. 1728-1734, 2007.

WILLIAMS, D. L. et al. The use of a specific DNA probe and polymerase chain reaction for the detection of Mycobacterium leprae. **The Journal of infectious diseases**, v. 162, n. 1, p. 193-200, 1990.

WORLD HEALTH ORGANIZATION- WHO. Available at: < www.who.int>. Accessed on: December 20, 2013.

WOODS, S. A.; COLE, S. T. A family of dispersed repeats in Mycobacterium leprae. **Molecular microbiology**, v. 4, n. 10, p. 1745-1751, 1990.

XING, Y. et al. VNTR typing studies of Mycobacterium leprae in China: assessment of methods and stability of markers during treatment. **Leprosy review**, v. 80, n. 3, p. 261-71, set. 2009.

YOON, K. H. et al. Evaluation of polymerase chain reaction amplification of Mycobacterium leprae-specific repetitive sequence in biopsy specimens from leprosy patients. **Journal of clinical microbiology**, v. 31, n. 4, p. 895-899, 1993.

YOUNG, S. K. et al. Microsatellite mapping of Mycobacterium leprae populations in infected humans. **Journal of clinical microbiology**, v. 42, n. 11, p. 4931-4936, 2004.

ZHANG, L.; BUDIAWAN, T.; MATSUOKA, M. Diversity of Potential Short Tandem Repeats in Mycobacterium leprae and Application for Molecular Typing. **Journal of clinical microbiology**, v. 43, n.10, p. 5221-5229, 2005.

ZODPEY, S. P.; AMBADEKAR, N. N.; THAKUR, A. Effectiveness of Bacillus Calmette Guerin (BCG) vaccination in the prevention of leprosy: A population-based case-control study in Yavatmal District, India. **Public Health**, v. 119, n. 3, p. 209-216, 2005.

APPENDIX A - QUESTIONNAIRE

1. Case or control: 1.( )Case 2.( )Control.SecontrolpuleforQuestion7

Case: patients who fit the research subject profile (diagnosis of leprosy; age between 5 and 65 years).

Control: patients who do not meet the requirements for the research subject profile.

2. Type of patient:

1) MDT-U/Caso Novo

Treatment-naïve patient, newly enrolled in MDT-U (project already underway at D. Libânia, or just one of the following

2) MDT-U Under treatment

Patient already participating in the MDT-U project and undergoing treatment

3) Pure Neural

Patient with no skin lesions, only nerve manifestations.

4) Single Lesion (Paucibacillary)

Patient presenting only a single skin lesion that is discarded at MDT-U

5) Multibacillary (not MDT-U)

Multibacillary patient who is not participating in the MDT-U project

3. Code in MDT-U: | ____ | ___|___ ____||

This information is only requested for participants in the MDT-U project. Enter the identification code assigned to patients already participating in the MDT-U project. This information can be found on the patient's appointment card. If it is not MDT-U and is only a new case, enter the code 00 (two zeros).

4. Registration Number: |||||

Enter the patient's medical record number. This information can be found on the patient's appointment card

5. Date of the 1os Symptoms: __/__/__

Enter the most precise date possible on which the patient remembers the first symptoms appearing. The date should be entered in the following format 00/00/0000, with slashes.

6. Date of diagnosis: __/__/__

Enter the date of diagnosis. This information can be found on the patient's appointment card. The date must be entered in the following format 00/00/0000, with the slashes being entered

7. Interviewer's letter and questionnaire number:

CASO: *Enter your Identifier ID (previously entered) and the number of the survey you are taking.* The number of *the questionnaire corresponds to the number of interviews you have carried out, regardless of whether they have been completed. Ex: you are conducting your first interview and your Identifier is G. In this space, enter G01.*

CONTROL: The controls have different numbers, starting with 1001, 1002, 1003 etc. The numbers must fall within the range 1001 to 1259. To find out the number of your control interview, check the "Spreadsheet for data collection (CONTROL)" and check the number corresponding to the last interview recorded. Your interview will use the next number and so on.

8. Date of interview: __________ / _ /.

Confirm that the date displayed is correct. If not, enter the date correctly, in the format requested by the questionnaire (00/00/0000).

9.	How many people live in your house?	Nowadays, ___ people Ask the patient to inform you of the number of people currently living in their home. And 10 years ago, people Ask the patient to tell you the number of people who lived with them 10 years ago.	
10.	CASE Do you know or did you know someone who had leprosy before you were diagnosed?	Yes No **(skip to 19)** I don't know **(skip to 19)**	
	CONTROL Do you or did you know someone with leprosy? *It seeks to find out if the patient knew anyone with leprosy before their diagnosis.*		
11.	How many people?	people	

12.	Who was it? *Mark the people diagnosed with leprosy with whom the interviewee had contact.* *More than one option can be ticked in this item.* *If you ticked "other", tell us what kind of relationship the interviewee had with the company.*	Relative (with whom you live frequently) Relative (with whom you live infrequently) Neighbor Friend Acquaintance Work colleague Other *(Specify)*	1 2 3 4 5 6 -8
	people with leprosy		
13.	Did or does this person still live with you in the same house? *To check whether the interviewee had direct and frequent contact with the person with leprosy.*	Yes No No answer	1 2 -9
14.	When was your first contact with this person? *How long ago was the last contact with the person with leprosy (years, months or days)?*	Years, Months Days *In this option, you should try to fill in all the spaces (years, months, days).*	
15.	How long (adding up all the periods) did contact with this person who had leprosy last? *Insert the time the interviewee was in contact*	years months days *In this option, you should try to fill in all the spaces (years, months, days) for*	
	with the person with leprosy. If this contact occurred at intervals, report	*more precise information.*	

	the total sum of the contact intervals. *For example, John had contact with Mary (with leprosy) for 1 year in 1999 and another 2 years between 2001 and 2003. You should enter "3 years" of contact, the sum of the exposure periods.*		
16.	How often is this contact made? *Tick the frequency, as closely as possible, with which the interviewee had contact with people with leprosy.* *If you ticked "Other", please specify this type*	Daily A few days a week More than 15 days a month Less than 15 days a month Monthly or a little more A few times a semester More than half the days of the year Less than half the days of the year Annually or a little more Other	
	contact details in the "Enter the relationship between this person" box that appears right after it.		
17.	What kind of contact did you have with this person(s)? *Inform the type of contact that the interviewed patient had with people with leprosy. If the "other contact" option is checked, state the type of contact.* *CONTROL interviewees skip to question 20.*	Slept in the same bed/network 1. Yes 2. No Slept in the same room 1. Yes 2. No They wore the same clothes 1. Yes 2. No Used the same bath towel 1. Yes 2. No Took care of her (bath, food) 1. Yes 2. No Was dating1 .Yes2 . No Worked together1 . Yes2 .No Other contact1 . Yes2 . No Specify__________________	
18.	How do you think you got leprosy?		

	Seek out the hypothesis of the patient interviewed (CASO) about how they became infected. Formulate a concise answer according to		
	report. *This question will not be asked of CONTROL interviewees.*		
19.	When do you think you caught leprosy? *Try to establish a date with the patient when they think they contracted leprosy. Seeking a date is the highest level of accuracy possible. Although patients may not remember the exact date, they often associate it with some event, such as the onset of symptoms, or even stipulate the period according to the information provided by doctors ("You must have contracted the disease a couple of years ago, etc.').*		
	This question will not be asked of the CONTROL interviewees.		
20.	CASE Had you heard of leprosy before your diagnosis? CONTROL Have you ever heard of leprosy? *Check whether the patient had any prior knowledge of the symptoms, types of manifestation, treatment, transmission or any other information*	Yes No	1 2

	about leprosy.		
21.	How old are you? *Inform the patient's age.*	years	
	In the case of a case subject, check that he or she is in the age group of inelusion in the research (between 5 and 65 years old).		
22.	Which sex? *Select the patient's gender*	Male Female	1 2
23.	Skin color *Ask the patient what color they consider their skin to be*	White Brown Black Yellow No answer	1 2 3 4 -9
24.	What is your marital status? *Ask the patient what their marital status is.*	Single Married/J u nto/loved Divorced/Divorced/separated Widowed No answer	1 2 3 4 -9
25.	What grade did you go to? *Ask the patient about their level of education. If they can't remember, try to work it out with them.*	I didn't go to school First grade incomplete First grade complete Second grade incomplete Completed high school Superior Higher education incomplete	1 2 3 4 5 6 7 8

		Postgraduate studies No answer	-9
26.	Which of these income brackets does your family fall into? *Calculate together with the patient how many minimum wages the members of the family earn and mark which income bracket the family falls into.* *The current value of the minimum wage is R$415.00.*	No income Less than one minimum wage 1 to <4 minimum wages 4 to <8 minimum wages 8 to <12 minimum wages 12 or more minimum wages Don't know No answer	1 2 3 4 5 -3 -9
27.	How educated is the head of your family? *Check the level of education of the head of the household. The head of the household may be the patient interviewed or a relative, usually identified as the head of the household.*	I didn't go to school First grade incomplete First grade complete Second grade incomplete Completed high school Higher education incomplete Higher education Post-graduation	1 2 3 4 5 6 7 8 -9
	o *financial provider*	No answer	
28.	How many of these items do you have in your home? Cars *Consider como automobile seed cars. Do not include motorcycles.* Bathroom *Include the number of bathrooms the person reports having in their home.*	*For all the items mentioned, you can mark the quantities described below (one, two, three or more than four items). There are also the options "Don't know" (NS), ticked if the patient doesn't know how many items are in their home, or "No answer" (NR), if the patient doesn't want to answer the question.* 01234+ NS NR	

	Monthly employee *Enter the number of employees who work daily and receive a monthly salary.* Freezer or duplex refrigerator. *Enter the sum of the number of freezers and duplex refrigerators, popularly known as "two-door refrigerators";* Simple fridge *Inform o number of fridges com internal freezer compartment*	01234+NSNR 01234+NSNR 01234+NSNR	
	Washing machine *Enter the number of washing machines* Radio *Enter the number of radios that the patient interviewed claims to own.* Color TV *Enter the number of color televisions.* VCR/DVD *Inform the sum of the number of VCRs and DVDs that the interviewed patient claims to own.*	01234+NSNR 01234+NSNR 01234+NSNR 01234+NSNR 01234+NSNR	
29.	Have you always had enough food for your family? *Ask the person if they think they have always had food enough for the whole family; if it's "passed fame".*	Yes No I don't know No answer	1 2 -3 -9
30.	Have you ever experienced major financial difficulties in your life?	Yes No I don't know No answer	1 2

		-3 -9
31.	Have you taken BCG? *BCG is the vaccine used to immunize against tuberculosis. It can leave a small scar at the application site.*	Yes No I don't know No answer
32.	Presence of a scar on the right arm? If *there is a scar as a result of applying BCG*	Yes No Not verified
33.	Do you smoke?	Yes No (Skip to 36) No answer (skip to 36)
34.	If so, how many cigarettes/day?	< 1 pack/day 1 pack/day More than 1 pack/day
35.	How often do you consume drinks alcohol?	Never (skip to o item "city of birth") Once a month or less 2 to 4 times a month 2 to 3 times a week 4 or more times a week
36.	How many doses of alcohol do you consume on a normal day? *A normal day means the days on which the person usually drinks, whether during the week or at weekends* *It's worth noting that a shot of alcohol means: 1 can of beer; 1 shot of*	O or 1 dose 2 or 3 doses 4 or 5 doses 6 or 7 doses 8 doses or more

	cognac or whisky; 1 glass of wine; 1 shot of aperitif; 1 glass of pinga, cachaça or caipirinha).	
37.	How often do you consume five or more doses on a single occasion? *Report the frequency with which the patient ingests five or more doses numa a single consumption occasion*	Never Less than once a month Once a month Once a week Almost every day

34. PLACE OF BIRTH

City:State: |_|_|

35. LIFE HISTORY

1) places where he lived

ADDRESS WITH 0 MAXIMUM INFORMATION FROM MOST RECENT TO OLDEST

Inform the addresses where the patient lived, from the most recent period to the oldest.

It is very important that the addresses recorded are as accurate as possible. If it is impossible for the patient to remember the number of the residence, inform them of an intersection between streets near where they used to live and a known point of reference.

The use of mapa is very useful, because by visualizing an avenue near where they lived, the interviewees were often able to locate no mapa the street where they lived.

As patients often get confused about the chronological order of the addresses they have lived at, it is interesting to try to organize the addresses first on paper, checking with the patient that the order is correct, and then answering the questions on the PocketPc.

M1 *Period (years)* *Put the year you started and the year you finished living at this address*	City	Neighborhood *Use o mapa, if the patient doesn't remember the neighborhood.*

		Address: ______________________________	№
		Complements: ______________ Reference point: __________	
		Inform nearby streets or points of reference, so that specific coordinates can be assigned to the places of residence.	

2) PLACES WHERE YOU STUDIED

This section should include the schools that the interviewee attended during regular schooling, as well as institutions for vocational courses, foreign languages, etc. When filling in the questions, the chronological order of courses should be followed, from the most recent to the oldest.

In the case of regular education (primary, secondary and higher education), check that the total number of years in the above-mentioned series matches the total number of years of schooling.

It is important to provide the address with as much information and as accurately as possible, so that specific geographic coordinates can be established for each study site. The use of mapa helps to find neighborhoods and streets.

If the interviewee seems confused about the chronological order of the places where they studied, they can write down the information on paper, organize it together with the patient and then answer the questions on PocketPC.

E1 *Period (years)* *Enter the year in which the study period began and ended.*	City	Neighborhood
at this location		

		School name
		Address: *Inform the address as precisely as possible. If the patient can't remember the name of the street, even with the aid of a map, insert places of reference, names of major avenues or crossroads where the school was located.*
		Address:

		Address:

3) PLACES WHERE YOU HAVE WORKED

List from most recent to oldest, filling in the address field with as much information as possible. It's also a good idea to write down in advance the different places where the interviewee worked, to help organize them chronologically.

What year did you start working? |||||

Inform the year in which the interviewee first worked

Il Period *Enter the year in which the interviewee joined and the year in which they left the company.*		City	Neighborhood
		Company or type of activity: *Inform o the name of the company or o the type of activity to which the company was dedicated*	
		Address: *Enter the most reliable address possible, with the help of a map. If the patient can't remember, ask about landmarks and nearby avenues or streets so that geographic coordinates can be established later.*	

4) LEISURE FACILITIES

Places you've been to in the last 10 years (address com maximum information)

Report the four leisure activities that the patient does most often. If the interviewee does the same activity in different places, record the places they usually go to do that activity.

E.g.: The interviewee chose "Beach" as his first choice of leisure activity. This is due to the beach huts that he used to frequent.

Here are some suggestions:

1. Beach
2. Cinema
3. Bathing or swimming in Iagoas, rivers, cacimba or other collections of water
4. Visiting relatives or friends
5. Talking to neighbors on the sidewalk
6. Visiting people and/or relatives who are in prison
7. Fishing
8. Hunting
9. Walk/Run/Cooper
10. Sport (soccer, volleyball, basketball, etc.)
11. Lining or dance venues
12. Bars
13. Church
14. Clubs
15. Shopping (mall, supermarket, bakery, butcher, bank)
16. Other

<table>
<tr><td rowspan="3"><u>L1</u></td><td>City</td><td>Neighborhood</td></tr>
<tr><td colspan="2">Location:
Name of the place where the person performs a certain activity Customary</td></tr>
<tr><td colspan="2">Address:
The most dignified possible address of the place where the activity takes place. If the patient can't remember the address, insert reference points so that geographical coordinates can be entered later.</td></tr>
</table>

36. DESCRIPTION OF ACTIVITIES - WORK/STUDY/LEISURE (IN RELATION TO THE LAST 30 DAYS)

If the activities listed in the previous questions, including work, study and leisure activities, have been carried out in the last 30 days, the following questions should be asked.

Length of	Area	Number of	Frequency of	Routine before/after diagnosis	Length of time in

stay (hours/day) *How long that activity lasts on a daily basis*	1 Open 2 Closed	people you share the activity area with per day	activity per day of the week *How many times a week does the interviewee do that activity?*	Indicate, according to the options, whether or not the interviewee's routine changed after the diagnosis. *This question will not be asked of CONTROL interviewees.* *1* before diagnosis *Check this box if the patient carried out seed activity before the diagnosis; if after the diagnosis, the patient stopped this type of activity.* *2* after diagnosis but before the end of treatment *Check this box if the activity started after diagnosis but before the end of leprosy treatment.*	the job YearsMonths Fill in the fields according to the length of time reported by the interviewee. E.g. If the person reports 10 months, fill in 0 in the "years" field and 10 in the "months" field. Proceed in the same way if the interviewee has been working for years.
				3. After discharge (end of treatment) *Check this box if the activity started after the end of treatment.* 4. Unchanged *Check this box if the activity began before the diagnosis and continues today.*	activity.

37. BASED ON THE TASKS IN THE TABLE ABOVE, WHAT TRANSPORTATION WAS USED TO CARRY THEM OUT?

Excerpt:	Type of	Transport	Are there	Travel	N^0 average	Time spent

1 Home - Locai 2 Other *In the case of the "other" option, describe any different route that the subject takes to carry out the activity*	transport 1 bus 2. Van/Topic 3 car 4 train 5 bicycle	line used (specify main line) *Only for buses and topics.*	any other lines used? 1 Yes 2 No *If yes, please specify*	time (min)	number of people who traveled by transport	using this transpcrtation Years Months
previously described	6 on foot 7 other		*which lines.*			

38. To carry out your work, you need

1. Staying in a fixed location 2. Moving from one place to another

---

39. Does your work involve contact with patients in healthcare environments (hospitals, health centers, doctor's offices)?

1. Yes2 . No

40. Have you ever been to prison?

In this case, the question is addressed to the patient who has already been in prison.

1. Yes2 . No

41. How many people visit you (or	How often you receive visitors	Visiting time (in hours)
frequent your home) per week ?	(days per week, month or year)	
	(days)/ (week, month, year)	

42. How many different places do you go to each month?

Estimate the number of different places the person goes to each month.

Frequency with which you receive visits Visiting time (in hours)

ANNEX A - OPINION OF THE ETHICS COMMITTEE

research ethics committee

DONA LIBÂNIA NATIONAL REFERENCE CENTER FOR DERMATOLOGY - CDERM

RESEARCH PROJECT ANALYSIS

Project Registration CEP/CDERM - 011/07

Title: Leprosy Patients: A Clinical-Epidemiological Study of Leprosy in an Area

Endemic to Northeast Brazil

Responsive researcher: Ligia Regina Franco Sansigolo Kerr

Proposal by the rapporteur:

(X) approval

( ) approval with reservations

( ) no approval

Date of first analysis by CEP/CDERM - 12/09/2007

Date of final opinion of the project by CEP/CDERM: 12/09/2007

OPINION

Based on CNS/MS Resolution No. 196/96, which regulates the matter, the Research Ethics Committee of the Dona Libania National Reference Center for Dermatology, CDERKL, at its meeting held on September 12, 2007, decided to APPROVE, in accordance with the Rapporteur's opinion, the research project specified above, in terms of its ethical aspects.

Comments:

1. We declare that Dr. Maria Araci Pontes, a member of this CEP, did not participate in the evaluation of the project in question and abstained from voting to avoid a conflict of interest.

2. The researcher(s) must submit a final research report to CEP/CDERM.

Fortaleza; September 13, 2007.

Dr. Lucy^teia Landim Tavares
Vice-Coordinator of CÊP/CDERM

Printed by Books on Demand GmbH, Norderstedt / Germany